PRAISE FOR *SPACE 2.0*

"Optimistic, but not over-the-top so. Comprehensive, from accurate history to clearly outlined future prospects. Sensitive to the emerging realities of the global space enterprise. Well-written and nicely illustrated. In *Space 2.0*, Rod Pyle has given us an extremely useful overview of what he calls 'a new space age.'"

—John Logsdon, professor emeritus at Space Policy
Institute, George Washington University

"The most exciting developments in the human exploration and development of space in the twenty-first century is the commercialization and internationalization of space access. New firms have emerged to provide cargo and soon human access to the International Space Station, and a broader range of international actors have developed capabilities only dreamed of twenty years ago. Rod Pyle tells this story with verve and style, offering a unique perspective on the second space age, *Space 2.0*, and its possibilities yet to be realized."

—Roger Launius, former chief historian for NASA

"*Space 2.0* offers a grand overview of everything happening in space from JPL's robots to Elon Musk's biggest ideas and that is saying a lot! Pyle makes the complex technologies and intractable policy debates behind all this accessible to any reader without dumbing them down. It's a great read for those who already excited about our new future in space and a must-read for those who do not yet get it. Buy one for yourself and two for loaning to your friends."

—Greg Autry, director of the University of Southern California's Commercial
Spaceflight Initiative and former NASA White House Liaison

"For spaceflight fanatics like me, now—the dawn of the Second Age of Space Exploration—is the most exciting time to be alive since the moon missions of the late 1960s and early '70s. But this complex new frontier of SpaceX, Virgin Galactic, space tourism, Mars cyclers, and asteroid mining is as different as could be from the Apollo era. In *Space 2.0*, ace science writer Rod Pyle—an advisor to NASA and the National Space Society—provides an engaging and expertly informed explanation of how we got this far, along with a factual yet inspiring intro to our around-the-corner

new adventures in space. Strap yourself in tight. It's a fascinating ride! Have spacesuit, will travel."

"*Space 2.0* is just the right book at just the right time. This important work not only provides a great historical background on space exploration, but more importantly, it highlights the next steps in the establishment of a deep space economy which includes, resource mining, additive manufacturing, infrastructure development, permanent habitation and eventually settlement. This is a must-read for anyone interested in the future of space exploration and development in the 21st century, and gives excellent suggestions about how you can get involved in *Space 2.0*."

"A clear and enthusiastic account the current and future era in human spaceflight, concise and accurate, yet packed with information. It's a primer on rocketry and other space-related industries, national and international space programs, key facilities, planetary defense, asteroid mining, bases in space and more. *Space 2.0* is a beginner's must-read for anyone contemplating an investment in the space industry or a career in spaceflight or space policy."

"As Elon Musk celebrates more than 50 successful launches and a plethora of successful landings of his Falcon rockets and as Jeff Bezos achieves the ninth successful launch and landing of his New Shepard rocket, the space game is about to change. Rapidly. Your indispensable guide to the new space race is Rod Pyle's *Space 2.0*."

SPACE 2.0

SPACE 2.0

How Private Spaceflight, a Resurgent NASA, and

International Partners Are Creating a New Space Age

ROD PYLE

National Space Society

BENBELLA

BenBella Books, Inc.
Dallas, TX

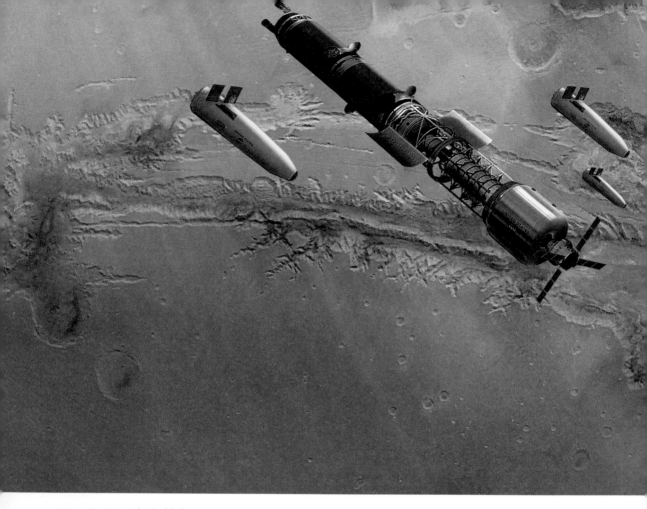

BENBELLA

BenBella Books, Inc.
10440 N. Central Expressway, Suite 800
Dallas, TX 75231
www.benbellabooks.com
Send feedback to feedback@benbellabooks.com

Printed in the United States of America
10 9 8 7 6 5 4 3 2 1

Library of Congress Cataloging-in-Publication Data is
available upon request.
9781944648459 (print)
9781946885746 (e-book)

Editing by James Lowder
Copyediting by Scott Calamar
Proofreading by Michael Fedison and Cape Cod
    Compositors, Inc.
Indexing by WordCo Indexing Services
Text design and composition by Aaron Edmiston
Cover design by Sarah Avinger
Cover illustration by James Vaughan
Printed by Versa Press

Distributed to the trade by Two Rivers Distribution, an
Ingram brand
www.tworiversdistribution.com

**Special discounts for bulk sales (minimum of 25 copies) are available.
Please contact bulkorders@benbellabooks.com.**

This book is dedicated to Stan Rosen and the members of the National Space Society, who have envisioned our future in space and worked tirelessly to ensure that future generations enjoy the bountiful returns of a true spacefaring civilization.

CONTENTS

Foreword by Buzz Aldrin   xiii

Chapter 1   1
Space 1.0, the Final Act

Chapter 2   15
A Dark and Forbidding Place

Chapter 3   25
Why Space?

Chapter 4   39
The First Space Age

Chapter 5   59
Destinations

Chapter 6   69
The Human Factor

Chapter 7   93
Space Entrepreneurs

Chapter 8   111
*Space Exploration Technologies
Corp.*

Chapter 9   135
A New Space Race

Chapter 10   151
Investing in Space

Chapter 11   163
The Space Between Nations

Chapter 12   183
The Russian Juggernaut

Chapter 13   191
The China Wildcard

Chapter 14   203
Truck Stops in Space

Chapter 15   223
Defending Earth

Chapter 16   233
Settling the Final Frontier

Chapter 17   255
A New Age Dawns

Chapter 18   263
Your Place in Space 2.0

Acknowledgments   273
Special Thanks To   276
Notes   277
Glossary   289
Index   294
About the Author   317
Join the National Space Society   319

Image credit: James Vaughan

FOREWORD

About a half century ago, in July of 1969, I participated in a great adventure that was the culmination of the first space age—the first humans to land on the moon. As Neil Armstrong and I climbed down to the desolate expanse of the Sea of Tranquility, with our crewmate Mike Collins orbiting overhead, we were cognizant of the vast, organized effort by an entire nation that delivered us safely to that distant body. Five more landings followed, and just three short years later, the Apollo lunar landing program came to an end. This was followed by Earth orbital missions through the space shuttle era and continuing with the International Space Station today, but in many ways, you could say that the first era of space exploration culminated with the return of Apollo 17's crew in 1972. We have not transported humans beyond Earth orbit since that time. And while much has been achieved since then by the United States, Russia, our partners in the ISS, and now China, it has all been conducted within a few hundred miles of the Earth's surface.

That is simply not good enough.

The promise of Apollo, felt not just by the astronauts but by people all over the world, was one of ever-greater expeditions to other worlds and in-depth explorations of them by human beings. Apollo was a profound beginning, but the implied promise of that first space age was to explore the moon with long-term habitations, then to reach out beyond to Mars and its moons. Plans were drawn up by both NASA and the Russians for increasingly ambitious human expeditions into deep space, but soon tucked away in file cabinets. Human spaceflight budgets were slashed, goals reined in, and justifications for these less aggressive programs were put forth. But for many people in all walks of life, myself included, it was an unsatisfying end to a decade of brilliant accomplishment. A bit of spark left the human spirit.

Since that time, I've remained active in space exploration efforts, sitting on countless committees and speaking at innumerable venues, in an attempt to drive this important goal forward. I've designed programs, such as semipermanent cycling transports to both the moon and Mars, that would allow for regular and affordable transportation. I've engaged the space community and laymen alike in an effort to enhance understanding of the seminal value of continuing our journey to the stars. I've also been involved with the National Space Society for many years, as a member, a leader, and a longtime supporter and active participant.

I am now increasingly optimistic that this new era of space exploration and development will dawn during my lifetime. And not just expeditions of exploration, important though they are, but for the long-term settlement of space by human beings. This is the nature of our species—to strive, to overcome limitations, and to move beyond our present boundaries. In the twenty-first century, this means moving human enterprise into space.

In the past few years, I have been delighted to see progressively stronger movement in the private sector toward this goal. A realm long thought to be primarily the domain of governments—low Earth orbit (commonly abbreviated as LEO) and beyond—is being accessed with increasing frequency by companies such as SpaceX, United Launch Alliance, and a growing range of international entities. Although these projects are still being funded to a large extent by government, within a few years I see private investment taking a leading role in these efforts. Internationally, China has joined the United States and Russia in human spaceflight, and India has taken an active role in planetary exploration. Others will follow.

This is music to the ears of any former moonwalker, along with millions of others who see a bright future in space.

We are witnessing the dawn of a new space age that will finally make traveling into space an affordable and profitable exercise—not just in monetary terms, but also in the full realization of the human spirit. Countless entities both inside national governments and in boardrooms—and with the advent of new technologies, even in university labs and residential garages—are taking aim at space, each according to their strengths and abilities. We will soon see a zone of convergence in which these parties will both compete and cooperate to regularize a human presence in space, with benefits for all of us. These benefits range from better agricultural practices that will feed additional billions, to limitless clean energy, to the expansion of human communities beyond Earth. The bounty to be reaped from a truly robust presence in space is impossible to overstate. An intelligently planned and consistent expansion into space holds promise for virtually everyone on Earth.

Please join me as we explore the next few decades in spaceflight in this important new book, produced in partnership with the National Space Society, and explore humanity's movement into the universe. It's time to get us off this planet, and to realize our destiny in space.

—Buzz Aldrin
Astronaut and global
space statesman
National Space Society
Board of Governors

CHAPTER 1

SPACE 1.0,
THE FINAL ACT

"If you want to build a ship, don't drum up people together to collect wood and don't assign them tasks and work, but rather teach them to long for the endless immensity of the sea."
—Antoine de Saint-Exupéry, author and pioneering aviator

"When something is important enough, you do it even if the odds are not in your favor."
—Elon Musk, founder of SpaceX

July 8, 2011, could have been any other summer day on the Florida coast—the sky was a stunning blue, a warm breeze blew across the tidelands, and the green water offshore was glassy. But this was not just another day in the Sunshine State. Before me was a billboard-sized digital clock, slowly counting down toward zero, and three miles beyond that stood a space shuttle, shimmering in the midday heat. Billows of white vapor spewed from its orange fuel tank as the ultra-cold liquid oxygen boiled off, as if

1

The space shuttle Atlantis *is rolled out for the program's 135th and final mission in 2011. Image credit: NASA*

up-and-coming NASA partnership with commercial space providers SpaceX and Boeing, who were building the presumed replacements for the shuttle's delivery of astronauts to the International Space Station (ISS). The wistfulness was for the end of the first era of human spaceflight, Space 1.0, which began with the early Mercury program and culminated in the Space Transportation System, more commonly known as the space shuttle. The shuttle program had been America's premiere space effort for over thirty years. It had returned us to space in 1981 after the six-year post-Apollo lull (the last Apollo spacecraft flew in 1975), conducted countless experiments in orbit, deployed and repaired the Hubble Space Telescope, docked repeatedly with the Soviet Mir space station, and built the International Space Station. While the shuttle never fully realized its intended destiny as a more affordable and reusable means of reaching space than its predecessor—the shuttle was complex, delicate, dangerous, and expensive to refurbish—the fleet served NASA for more than three decades, providing relatively routine access to low Earth orbit. The July 2011 launch represented the final act of the program.

Atlantis's primary payload that day was a Europe-supplied component called the Multi-Purpose Logistics Module. *Atlantis* carried only four astronauts instead of the normal six or seven, due to the fact that

anxious to be ignited. *Atlantis* appeared ready to bolt into the heavens.

This morning would host the 135th and final launch of the space shuttle program, an exciting event, as all launches are, but also one swaddled in both hope and wistfulness. The hope hinged on the

Tens of thousands gather on the north side of the Kennedy Space Center near Titusville. Image credit: NASA

Atlantis lifts off at 11:29 AM Eastern Time, July 8, 2011. Image credit: NASA

there were no more shuttles available for a rescuc in case of an in-flight emergency. If a problem occurred, the astronauts would have to take shelter in the ISS, and four was the maximum number that could be safely accommodated there.

As the countdown reached its final seconds, a gush of water erupted at the base of the launchpad to absorb the massive acoustic shock from the rocket engines that would soon pummel the flame trench. A bright flicker emanated from the base of the shuttle, and a tremendous volume of smoke billowed outward as the engines came up to full power. Moments later, at T minus zero, the side-mounted solid rocket boosters ignited and *Atlantis* leapt skyward, departing on its final journey at 11:29 AM. A deep rumble washed over us a few seconds later as the spacecraft climbed slowly skyward, gaining speed and riding a column of smoke into the east.

During a previous launch, the shuttle Endeavor lifts clear of cloud cover. Image credit: NASA

Atlantis departs the cape amidst fire and thunder. Image credit: NASA.

As the smiling crowd slowly departed the stands, a sense of finality settled over me. Nobody knew for certain when another rocket would depart from Pad 39. America's future in human spaceflight was dependent on new partnerships with private aerospace companies, an arrangement that was in flux. It could be years before US astronauts departed from the cape again.

As the spectators headed home, *Atlantis* made its way to the space station. Before docking, it paused about six hundred feet away for astronauts already aboard the ISS to give the shuttle's tile-covered underbelly and the leading edges of its wings a thorough inspection. Those tiles, and the reinforced carbon composite shells that made up the front ends of the wings, would be its sole protection from the incandescent heat of reentry it would face in just under two weeks. NASA had learned the hard way that damage to these elements of the thermal protection system could result in catastrophe, as had occurred eight years earlier with the shuttle *Columbia*. A piece of foam insulation from the fuel tank had broken free during launch and impacted the leading edge of the left wing, punching through the thin reinforced carbon composite there, dooming *Columbia* and its crew. The *Columbia* accident was the second loss of a shuttle since *Challenger* exploded in 1986, and just one more in

Atlantis *orbits below the International Space Station as it prepares to dock. Image credit: NASA*

Atlantis *flies belly-up near the International Space Station's windows to allow ISS astronauts to make an assessment of the heat-shielding tiles. Image credit: NASA*

a long list of flights in which protective thermal coverings were damaged during launch, imperiling the crew. This fragility was one of a number of technical factors that led to the termination of the shuttle program.

With that solemn assessment complete, *Atlantis* docked with the ISS. Then the shuttle's mission continued for its twelve-day duration. *Atlantis* returned home for the last time on July 21, 2011, was decommissioned, and now resides in a museum display at the Kennedy Space Center, locked in a permanent orbit around a huge rear-projected image of the earth.

The end of the shuttle program created a yawning gap in America's ability to fly humans into space. The Constellation program, initiated by President George W. Bush in 2004, was intended to create a replacement for the shuttle—a cheaper, expendable rocket and capsule, designed along the lines of the Apollo spacecraft of the 1960s. This new program was also supposed to have the capability to return astronauts to the surface of the moon. But Constellation, continually over schedule and underbudgeted, was cancelled by President Obama in 2010 after a program review by aerospace experts, so the shuttle program came to an end without a replacement.

Let me repeat that: America was giving up its sole functioning capability to fly astronauts to the space station that we had designed and built, at staggering expense, without a domestic replacement. This

A view from the shuttle's cockpit as Atlantis glides to its final touchdown at the Kennedy Space Center. Image credit: NASA

Atlantis *seen after landing. Image credit: NASA/Pete Crow*

stunned much of the spaceflight community, and a lot of people, including a number of Apollo-era astronauts, were quite vocal in their dissent. Neil Armstrong, the

Atlantis on permanent display at the Kennedy Space Center, with a permanent orbital sunrise in the background. Image credit: NASA

first man to walk on the moon; Jim Lovell of *Apollo 13* fame; and Gene Cernan, the last man to depart the lunar surface, all strongly condemned the decision. In an open letter to the president, the three former astronauts explained that the choice to cancel the Constellation program was "devastating." The letter continued: "America's only path to low Earth orbit and the International Space Station will now be subject to an agreement with Russia to purchase space on its Soyuz . . . until we have the capacity to provide transportation for ourselves. The availability of a commercial transport to orbit as envisioned in the president's proposal cannot be predicted with any certainty, but is likely to take substantially longer and be more expensive than we would hope."[1] Their assessment of the pace at which commercial spaceflight could develop alternatives to deliver astronauts to orbit turned out to be prophetic. While much

progress has been made since that final shuttle launch, many experts suggest that commercial spaceflight partnerships have been consistently underfunded, and the contractors that will be providing astronaut transportation to the ISS—Boeing and Elon Musk's SpaceX—have been running behind schedule. While nobody can say how the Constellation program would have turned out, the US has suffered a significant gap in human spaceflight capability since the end of the shuttle program.

The shuttle's temporary replacement, the Russian Soyuz, was to cost NASA $32 million per seat for each American astronaut that flew to the ISS. That sounds expensive, but was on its surface much cheaper than flying on the shuttle, which cost a *minimum* of $71 million per person. This represented a large cost savings over the development of the Constellation program . . . if low Earth orbit were your only destination. But Constellation

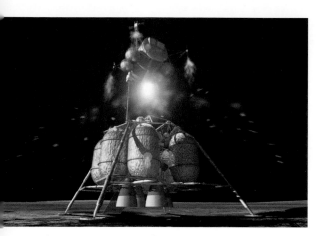

The ascent stage of NASA's never-realized Altair lunar lander departs to return to an orbital rendezvous with the Orion spacecraft, currently under construction by the agency. Altair was part of the Constellation program, cancelled in 2010. Image credit: NASA

Neil Armstrong (left) confers with Gene Cernan (right) during testimony about the cancellation of the Constellation moon program in 2010. They were the first and last men on the moon. Image credit: NASA

promised the ability to send missions to the moon and eventually Mars, so the numbers are difficult to compare. Constellation would also have been capable of delivering much larger, heavier payloads to orbit. In the short term, buying seats on Soyuz seemed like a bargain, but it was not. By 2017 that price had risen to $81.7 million, arguably more per passenger than the shuttle, and with far less capability.[2] Not such a bargain anymore.

There are a number of reasons for this discomfort with the Soyuz as a delivery vehicle besides the direct per-person launch costs. While the Russian launches were (initially) cheaper, the use of the Soyuz restricted American astronauts to Russian flight schedules, which were often erratic. US astronauts would have to train at Russian space facilities, which represented an added expense. The Soyuz capsules were also much smaller than the shuttle's crew compartment, carrying a maximum of three astronauts as opposed to the shuttle's complement of seven. Finally, manned Soyuz flights were unable to carry a meaningful amount of cargo to the station, so the US would end up paying for robotic cargo flights to offset what the shuttle was no longer delivering.

There was an additional objection contained in the letter from the three astronauts that went beyond complaints about Russian pricing and the uncertain future of commercial spaceflight, one more philosophical in nature. They continued: "Without the skill and experience that actual spacecraft operation provides, the US is far too likely to be on a long downhill slide to mediocrity. America must decide if it wishes to remain a leader in space. If it does, we should institute a program which will give us the very best chance of achieving that goal."

Their contention was that the US, a premiere power in spaceflight, was ceding its ability to get people into space and, perhaps more critically, its ability to operate large human spaceflight programs in the future. In response, the Obama administration outlined plans to spend more money to support development of human-rated spacecraft in the private sector, and to salvage parts of the Constellation program—an evolution of Constellation's heavy booster, now dubbed the Space Launch System (SLS), and the Orion capsule.

To many observers, this represented a welcome change in priorities. NASA would continue to build—albeit very slowly—its own space capsule and large rocket, which would carry a human crew to rendezvous with an asteroid in the 2020s. NASA would additionally direct funding to private spaceflight companies such as SpaceX and Boeing to deliver astronauts and supplies to the ISS.

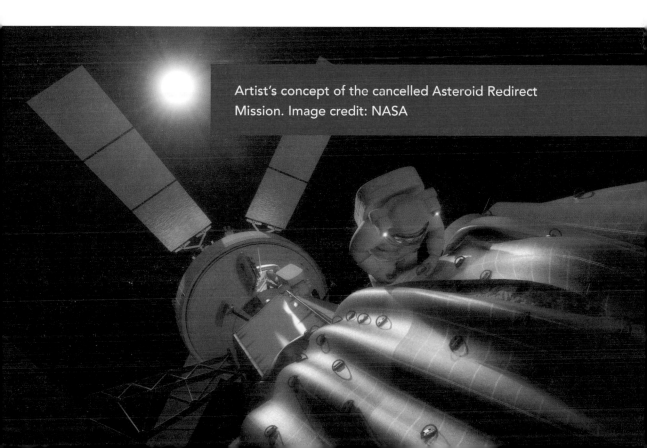

Artist's concept of the cancelled Asteroid Redirect Mission. Image credit: NASA

There was one major bug in the oint-ment, however. The Asteroid Redirect Mission (ARM), as it came to be known, while a valid demonstration of technol-ogy and engineering, was not popular. It appeared on the surface to be a program invented more to give Orion/SLS some-thing to do rather than being a valid sci-ence and engineering program. ARM died with the start of the Trump admin-istration and has since been replaced with a NASA initiative to return to the moon.

This is where we find ourselves near the end of the second decade of the twenty-first century. Regardless of the destination planned for Orion/SLS, both vehicles have been consistently behind schedule and over budget. But the Orion/SLS system continues to move forward and is currently slated for a maiden flight in 2020. The realpolitik of the programs involves a great number of jobs in a few NASA facilities and aero-space contractors in various congressio-nal districts—always important political considerations.[3]

At the same time, NASA funding for private sector efforts has fallen short of its own stated goals most years since 2010. SpaceX, Boeing, and other com-mercial space providers continue to vie for increased government funding for commercial spaceflight capability. NASA's overarching plan is for the commercial providers to handle transporting people

President Donald Trump signs a bill reactivating the National Space Council in June 2017. Image credit: NASA

and cargo to the International Space Sta-tion, while Orion/SLS handles NASA's human exploration plans beyond low Earth orbit.[4]

This scenario may sound a bit bifur-cated, with too little money being spent in two connected camps, but there are reasons for optimism. In early 2017, the Trump administration reactivated a government oversight group called the National Space Council. This body had existed twice before—once between 1958 and 1973, then again from 1989 to 1993. Its mandate is to tackle long-range poli-cies for American goals in space, reporting directly to the executive branch, with the plan then being implemented by NASA and other federal departments. The coun-cil is chaired by the vice president, and

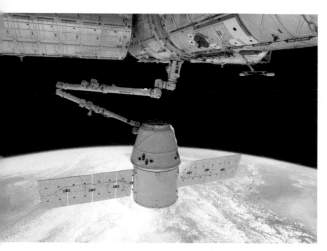

SpaceX's Dragon 2 capsule should begin ferrying astronauts to the ISS in 2019.
Image credit: NASA

©2016 Sierra Nevada Corporation

Sierra Nevada's Dream Chaser is a winged minishuttle that recently won preliminary approval by NASA to deliver cargo via commercial providers to the ISS.
Image credit: NASA

it's composed of top space policy makers, cabinet-level government officials, and the NASA administrator. It is advised by a selection of aerospace professionals from companies old and new.

The council has created recommendations for how NASA can best move ahead, both with its own efforts and with continued funding for the private sector. The preferred balance it seeks is for the private sector companies to be properly funded to do what they can do best—create a set of robust, low-cost spaceflight capabilities to get humans and cargo into orbit, and to compete among themselves to improve these services for increasingly lower costs.

The two companies currently developing a contracted capability to deliver astronauts to the ISS are SpaceX and Boeing. SpaceX is building and flying the Falcon 9 rocket, which is already running regular cargo supply missions to the ISS and is poised to begin flying astronauts to the station in 2019 or 2020. It has independently developed and flown the Falcon Heavy megabooster, but this is not, as yet, part of NASA's commercial program.

Boeing and United Launch Alliance (ULA) have their own contracts with NASA, and a Boeing-built capsule called the Starliner will soon be launching atop ULA's Atlas rockets to deliver crews to the ISS. These traditional aerospace companies know that they must adapt or die

Boeing's Starliner spacecraft, scheduled to begin crew-delivery service to the International Space Station in 2019. Image credit: NASA

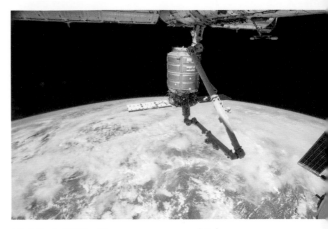

Orbital ATK's Cygnus cargo vehicle prepares to berth with the International Space Station after being grappled by the Canadarm, the ISS's robotic manipulator arm. Image credit: NASA

in this new environment and are working hard to compete with SpaceX.

Northrop Grumman Innovation Systems (via the acquisition of Orbital ATK), among other commercial entities, flies cargo missions to the ISS under contract to NASA. Sierra Nevada Corporation, a company that is experimenting with a mini-space shuttle, is preparing to join in on these cargo runs.

For decades, the US and Russia were the only countries flying humans into space, but this is changing. China launched the first crewed mission of its human spaceflight program in 2003 and has lofted two small space stations of its own. India is poised to follow with human spaceflight capability by 2022. Clearly, new international players and the

US commercial sector are gaining traction in an area that had been, since 1961, the exclusive domain of government agencies within the two major Cold War superpowers.

These developments are heartening, as they represent a strong shift in favor of more frequent, less expensive, and more diversified access for humans to travel into space. But back in 2011, as I watched the smoky trail of *Atlantis* dissipate over the Kennedy Space Center, these were still mostly tantalizing promises. SpaceX would not make its first cargo run to the ISS until 2012, Boeing was just beginning work on the Starliner, and China was striving to deliver the first crew to its tiny Tiangong-1 orbiting laboratory. These were hopeful beginnings, but it was, at the time, far from certain how any of these efforts would play out.

As I traveled home from Florida, I reflected upon the circuitous route that US space activities and those of other nations had taken since the 1960s. I thought about how the Apollo spacecraft had been replaced by the shuttle, ending our efforts for human exploration beyond Earth orbit. I reflected on what alternative paths might have been followed, and what the results might have been. I balanced these notions with positive thoughts about the rise of private spaceflight and China's nascent efforts. One thing became increasingly clear as I pondered the future: Whatever the next few decades held in store, the results were unlikely to resemble the first space age, and that was a good thing.

Since the late 1970s, many space advocates have pined for another "Kennedy moment," when an American president or foreign leader (probably Russian) would stride to a podium and announce a daring, large-scale government program to accomplish aggressive new goals in space, such as JFK did when he announced the Apollo lunar landing program in 1961. Three US presidents have made similar announcements, but none have stuck like the lunar landing challenge did—and little wonder. Kennedy's announcement, which kicked off the space race, was

Astronaut Gene Cernan adjusts the American flag during the final mission on the moon, Apollo 17. Image credit: NASA

motivated primarily by geopolitics and was very much a product of its time. It is not a moment likely to be reenacted in the twenty-first century; the world has moved on from the Cold War tensions of binary superpowers. We are now in a far better position to embark upon a sustainable program of human spaceflight and space development, and toward a permanent, robust human presence off Earth that will offer wide-ranging benefits for everyone who participates, directly or indirectly.

Welcome to Space 2.0.

CHAPTER 2

A DARK AND FORBIDDING PLACE

Space hates people. Don't let that put you off, though—well over five hundred humans have traveled its reaches, and most have returned safely.[5] These brave individuals flew because they felt it was worth it, despite knowing the substantial risks involved. I've never met an astronaut who would not willingly accept a reasonable amount of danger to accomplish a mission. Space is not a place that naturally lends itself to human travel and habitation, however, and much must be done to get people there and keep them safe, even for short periods of time.

Let's look at launch first. Why is getting into orbit such a challenge? Space is just overhead, and the unofficial boundary is only sixty-two miles above us. The International Space Station orbits every ninety minutes at an altitude of just 250 miles. Without the snarl of city traffic, you could drive that far in about four hours. But, as it turns out, going *up* as opposed to *across* is much more difficult—and to reach orbit, you must go extremely fast. That requires a lot more energy than moving across Earth's surface, and harnessing that energy is tricky business.

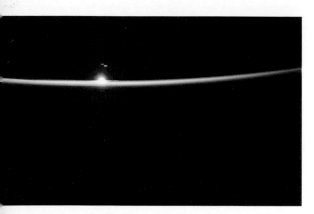

Sunset as seen from the space shuttle. Image credit: NASA

A long-exposure image of the shuttle taking off and heading eastward. Image credit: NASA.

When carrying spacecraft into orbit, rockets must aim from the surface of the earth in a curving path high above the equator, and then apply a lot of force to overcome the clutches of gravity to travel high enough, and fast enough, to reach and stay in orbit. So while you might need only ten gallons of fuel to make our 250-mile drive on Earth, traveling at sixty miles per hour for a little over four hours to get there, to reach the ISS you must not only travel to the 250-mile altitude required, but you must do so along a long, arcing line from the surface, usually aiming for a point in the sky well east of where you began.[6] By the time you get there, you must be moving fast enough to continue counteracting the force of gravity, which is trying its best to haul you back down to Earth. So besides the force required just to leave the Earth's surface (which even a Cessna airplane can do, after all), one must ultimately be traveling at a speed of about 17,500 miles per hour to reach and stay in orbit. That's why rockets are so large compared to the tiny capsules they propel into space; it takes that much fuel to overcome a standing start from the ground and reach the required speed and altitude.

About 90 percent of a rocket's mass is made up of fuel, leaving only 10 percent to carry the payload (people, robots, supplies, etc.). This fact is commonly referred to as the "tyranny of the rocket equation," which pertains to a derivation first conceived in 1903 by a famous Russian scientist named Konstantin Tsiolkovsky.

In fact, Earth's gravity is so powerful that a journey to Mars, tens of millions of miles away, only requires about twice the fuel that simply getting into orbit does.

Of the 363-foot height of the Saturn V moon rocket, only the last thirteen feet came home, the tiny conical command module at the top (outlined in red) that carried three crew members. Image credit: NASA

This gives you some idea of the forces that must be overcome just to reach the lowest orbit around our planet.

Most people have an intuitive sense of this. We've all seen videos of thunderous rocket launches carrying satellites, capsules, or space shuttles into the heavens. All that storm and fury looks complex and expensive—and it is.

As is said often in the aerospace trade, and with appropriate gravitas: *Space is hard.*

Once in orbit, with the mission complete, the spacecraft must reenter the atmosphere to come home. It must shed enough of the energy that was applied to get it up to 17,500 mph (orbital speed) in order to succumb to the attraction of gravity. In effect, you are reversing the process that took you into orbit, slowing the spacecraft enough to be pulled back to Earth. This is accomplished by firing retrorockets—small braking thrusters—to cause the spacecraft to lose enough speed to dip back into the atmosphere. Then, as you fall into denser air (there is effectively no atmosphere in orbit), friction from air molecules slows your spacecraft. By the time you dip back below that sixty-two-mile limit, the atmosphere becomes a meaningful presence, dense enough to cause slowing and heat via friction. The gas molecules simply cannot move out of the way fast enough. If your spacecraft is unprotected, it will begin to

The retrorockets on a Mercury capsule, circa 1960. These braking thrusters were what slowed the capsule sufficiently to reenter Earth's atmosphere. Image credit: NASA

melt and eventually disintegrate. Most meteors that enter Earth's atmosphere never reach the surface; all that friction vaporizes them. So your spacecraft needs to be sufficiently tough, and adequately shielded, to withstand the structural stresses and extreme heat of returning.

Reentering spacecraft overcome this heat with shielding materials. These compounds resist heating and protect the spacecraft from temperatures that can reach thousands of degrees. The space shuttle's heat-absorbing tiles were rated at about 1,500 degrees Fahrenheit, and the Apollo capsules, which were built to return from the moon at speeds of about

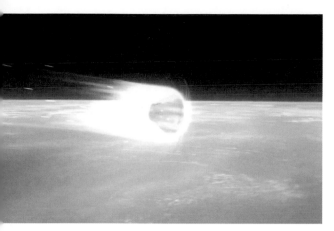

Artist's concept of an Apollo command module reentering Earth's atmosphere. Heating can rise to 5,000°F. Image credit: NASA

The space shuttle is seen reentering, with heat-resistant tiles aglow, in this 1970s NASA artwork. Image credit: NASA

25,000 mph, experienced heating up to 5,000°F.

After this scorching and buffeting by the atmosphere, other methods are used to continue slowing the spacecraft. Winged spaceplanes like the shuttle, and some new spaceplanes being fielded by private companies and the Air Force, use their wings to allow the craft to glide home, losing energy in the process and eventually descending to a runway. Space capsules, from the earliest Mercury craft to the most recent Soyuz and SpaceX's Dragon, use parachutes and sometimes braking rockets fired during the last moments of their descent to lose speed. Landing speeds range from about 200 mph for a spaceplane setting down on a runway (it sheds the rest of its energy as it

Aerial shot of a Soyuz capsule landing. The Russian spacecraft remains the only crewed capsule to perform landings on dry land. Image credit: NASA/Bill Ingalls

rolls to a stop) to about 7 mph for a Soyuz capsule or Dragon. In spaceflight terms, this is known as "delta-v," or a change in velocity. Going from 17,500 mph to just over 7 mph is a *lot* of delta-v.

When going to more distant destinations, such as the moon and, someday, Mars, return speeds are far faster due to the increasing acceleration the craft experiences as it races across space, greedily tugged by Earth's gravitational field. As with the Apollo lunar missions, more robust heat shielding than that used on the shuttles will be required to protect returning spacecraft.

These are the mechanics of getting to and from space. Keeping the human passengers alive during the trip up, their time in space, and the violent return to Earth is also quite difficult.

To understand this, it helps to think of people in more mechanical terms. In essence, we are water and meat contained inside a soft, squishy bag we call skin. We evolved from small, delicate, single-celled creatures into complex primates over almost four billion years, constantly challenged by the environment in which we were forced to struggle for survival. And that is the key: We were formed by the environment of the planet on which we originated. Earth eventually settled down from its toxic and rough-and-tumble origins into the relatively benign world we know today, and we are perfectly suited to it. A force of one Earth gravity is exerted upon us at all times—a product of your mass and the mass of the earth. Your body is also being squeezed by the atmosphere,

which at sea level is a force of about 14.5 pounds per square inch. Temperatures are generally somewhere between freezing at the polar regions to about 120°F in hot regions. We are most comfortable between 60 and 70 degrees; that's why most people in developed countries set their air-conditioning units, if they're fortunate enough to have one, within this range. We have evolved perfectly to exist in this environment.

To survive, we must drink water to keep our bodies hydrated, eat food to get energy to produce work (even if that work is merely keeping our hearts beating and blood flowing), and breathe air. Had we evolved on the moon or Mars, or for that matter on Venus, Jupiter, or even in space, things would have been much different.

Earth's atmosphere and gravitational environment are perfectly suited to human beings. Space is far less welcoming. Image credit: NASA

But we didn't, and we are stuck with the biological design that evolution gave us, here on the surface of a warm, wet planet with a blanket of air.

Space is an entirely different story. To survive there, we must carry a small portion of Earth's benevolent environment with us. The farther we travel into the cosmos, and the more time we spend there, the more rigorous our preparations must be. We must take air, water, and food to maintain life. During long missions, we must find ways to deal with the absence of gravity—our bodies don't do well without it over the long term. Meanwhile, the vacuum of space wants to suffocate us and pull the moisture out of our bodies, leaving us as dry, lifeless husks. The temperatures in the equivalent of our celestial neighborhood—say, between Venus and Mars—are tied directly to the nearby sun, and they are brutal. The side of an object in space facing the sun is quickly heated to about 250°F, and the shaded side plummets to about −250°F, and can go as low as −340°F. If unprotected in space, your body would quickly suffocate and lose its moisture. Then, one side would ice up into a solid, messy mass while the other would begin to bubble and cook.

These characteristics apply whether at the edge of space—about sixty-two miles from Earth's surface—or all the way out on the surface of the moon or in orbit around Mars. The surface of Mars is a

Radiation from the sun, and from beyond the solar system, is deflected by the Earth's magnetic field. Once spacecraft travel beyond Earth orbit, they are no longer protected. Image credit: NASA

bit more benign, but the temperatures still range from a high of about 70°F to a chilly −100°F at the equator. At the poles, it can plummet below −225°F. And the atmosphere is so very thin—less than 1/100th that of Earth's—that you would experience similar smothering and dehydrating effects on your body as you would in open space.

There is another demon living in space: radiation. If you move beyond the region known as low Earth orbit, and are properly protected from the vacuum and the temperature extremes, you must still deal with deadly radiation. Here on our home planet, and even in low Earth orbit, you are protected from most space-originated radiation by the planet's magnetic field, which redirects most incoming radiation to the poles. Our atmosphere further protects us by filtering out the ultraviolet radiation emitted from the sun. But once you get a few tens of thousand miles away from Earth, that protection disappears. In space, radiation from both the sun and other stellar bodies will cook your DNA over time, which ultimately results in various cancers, cardiovascular disease, and, ultimately, death. Even your brain may be affected by radiation over time, with confusion, depression, and a lack of focus as the result—all bad things when you are trying to operate a complex spacecraft.[7]

Finally, we must discuss gravity. We complain about hot and cold, and about strange smells in the air we breathe, or a lack of food, but gravity is simply always there, doing its thing, 24/7. Once we leave Earth and enter orbit, gravity is counterbalanced by centrifugal force—it ceases to have any meaningful effect on the body. This can become an issue for creatures that have evolved in a one-g environment. When that force is no longer being applied, all kinds of weird things happen. The fluids in your body, which normally pool in your legs and lower abdomen, equalize and drift back up into your chest and head. Astronauts who have spent any time in space say it's like having a permanent head cold. Then your bones, no

longer supporting your weight against the pull of the earth, shed calcium and become brittle. Your eyes change shape, gradually losing the ability to focus properly.[8] The effects of long-term partial-gravity environments, such as those found on the moon and Mars, may be less severe, but humans have not spent enough time in partial-gravity environments to gain an understanding of how they affect the body over extended periods. It may be that on Mars, which has .38 Earth's gravity, we will be fine over time. The moon has only about one-sixth of Earth's gravity. It is likely that either environment will be better than living long term in zero-g, but further research will be required to know for certain. We'll discuss this in more detail in Chapter 6.

We've been sending humans into space for almost sixty years, and in that time we have learned how to overcome some of these challenges, but not others. In the early years, flying small capsules in Earth orbit, we picked the low-hanging fruit of spaceflight technology. We learned how to use powerful chemical explosions, controlled and channeled through rocket nozzles, to propel people into orbit and beyond to the moon. We designed ever-improved space suits to protect us from the harsh vacuum of space and, later, from the hostile environment of the lunar surface. Since the space race years, we have learned a lot more about how to live in space, with a few individuals spending over a year in orbiting stations. But this is just the beginning of our long-term journey.

To reach the higher-hanging fruit will require learning how to live in space over the long term, to mediate the effects of low gravity and radiation. We will need to understand this better before extended spaceflight, and living in space, becomes routine. But we have made a solid start, and ongoing research in places like the ISS have taught us much. So while we are better at spaceflight than ever before, it is still difficult, risky, and challenging.

So an enormous question looms before us: If space is so hostile to humans, and so challenging to reach, why go?

The final moments of the Apollo 17 mission to the moon in 1972, the end of the opening act in the exploration of space. Image credit: NASA

Image credit: James Vaughan

CHAPTER 3

WHY SPACE?

I have been writing about space as an author and journalist for years. Part of this process is talking about space on radio, on television, and on the internet. A question that comes up again and again is why do we bother to leave our planet in the first place, and why are we striving to go farther, and stay longer, once we are there?

You probably have some thoughts of your own about why spaceflight and the long-term exploration of the cosmos is important, but there are plenty of people who do not share your vision. Explaining why you think spaceflight is important to others—people who may not be as well versed in the rewards of a robust space program and the resulting returns for humanity—can be a challenge. In fact, this is a topic that NASA has been addressing since its formation in the late 1950s. NASA, and those who support its mission, has struggled to convince both the Congress and the broader public why space exploration is important. As citizen-participants, sharpening our own thoughts on the subject can help us to explain this to others.

The first and possibly most essential question that often arises is this: What benefits does space exploration hold for the average person? Some of these radio and TV shows I participate in take calls and texts from the audience, and while many listeners are interested in space exploration, not everyone is convinced that it is important, and that's okay. Most of us involved in this discussion live in democracies, and debate is a part of the process.

Let me share a distillation of a typical conversation from some of the call-in shows I've been a guest on.

> Host: Our next caller is from Des Moines. Go ahead, Jeff.
>
> Jeff: Hi. I just wanted to ask your guest why we are spending all this money on space. I'm not sure I support NASA's expensive programs without something coming back for me. I mean, we have plenty of problems here on this planet. I can barely get across town in rush hour, and we still haven't cured cancer. What does the average person get out of it?
>
> Host: Okay, Jeff, thanks for the question. So, Rod, you heard Jeff. I'll add that we have already put men on the moon and we have a space station in orbit right now, but the tangible benefits seem scarce. Why should we continue the space program?

That's a valid question, and the host knows it—her audience wants to know why seemingly vast sums of their tax dollars are being spent on spaceflight instead of problems and challenges they consider more immediate.

Let's begin by correcting a few common misperceptions. First, there is no one, single US government "space program." NASA's civil space budget is a relatively small part of what the US government spends on space once you include military activities and other government agencies involved in weather prediction, climate research, and so forth. But NASA's budget is the single most visible chunk of federal space money, so public opinion on the usefulness of the US space program tends to focus on NASA. Polls consistently show that the average American grossly overestimates the amount of money NASA gets each year. The amount of the US budget allocated to NASA is just a tiny percentage—about one-half of 1 percent. That's about fifty-four dollars per year, per US citizen (at the height of the Apollo program, that cost was more like two hundred dollars, in adjusted dollars).[9] A 2007 study by Dittmar Associates, a Houston-based aerospace consultancy and think tank, summarized popular opinion regarding NASA's budget like this:

"Americans in general have no idea what NASA's 'cost' is. In fact, most members of the public have no idea how much any government agency's budget is." The study continued, noting that: "NASA's allocation, on average, was estimated to be approximately 24 percent of the national budget (the NASA allocation in 2007 was approximately 0.58 percent of the budget)."[10]

At its height in the mid-1960s, NASA's budget was almost 5 percent of federal dollars; it is now about a tenth of that. At the same time, the agency is running far more complex and wide-ranging programs than they were in the 1960s, including weather and Earth-science satellites, robotic planetary and deep-space exploration programs, and human-staffed efforts such as the ISS. The agency is also investing in private spaceflight. So NASA is, in fact, doing far more with much less.

In 1969 dollars, the Apollo program was estimated to have cost about $20 to $25 billion (or roughly $150 billion in today's dollars). Regardless of the total federal budget, that sounds like a lot of money. And to many people, it was too much to spend for what they thought of as just 840 pounds of moon rocks and a few thousand pictures. The shuttle program cost about $200 billion, and the ISS, while it was created by a consortium of nations, has cost a total of about $150 billion to date; about $76 billion of this

paid by the US.[11] NASA's human Mars program, when completed, is likely to cost the same amount or more—perhaps up to a trillion dollars over time.[12]

However, these figures should be examined in comparison to other technology-based federal programs to be relevant. The costs for the F-35 next-generation jet-fighter program currently stands at about $400 billion,[13] and while it's clear that the US needs to maintain its national security footing, the price tag puts NASA's projected space exploration budgets in a slightly different perspective. The US's 2019 defense budget is over $700 billion. NASA's annual budget of about $19 billion, when seen through this lens, does not seem large in comparison.

Moving ahead, if we accept that NASA's roughly $19 billion budget is reasonable, what do we get for our investment? The answer to this question is a long one and will unfold throughout the book, but in short: We get vast scientific returns from our human and robotic spaceflight programs. The discoveries and data from these programs continue to answer questions on the origins of our solar system (and the universe itself), the moon's composition (and how that might be useful to future space exploration), and future weather patterns (critical for agriculture and human health). The space program has helped us address a number of human health issues, from urgent questions about tissue regeneration (the healing of wounds), the loss of bone strength and its prevention on Earth as well as in space, the function of the inner ear and its role in balance and mobility, and immune system responses to stressors. Ground-based medical research to support human spaceflight has also yielded huge benefits, including better preventative medicine, improved detection and monitoring of diseases—there have been numerous advances in cancer research accomplished through spaceflight medical research—as well as advances in radiology and improvements in prosthetic limbs, to name just a few.

America's space investment is responsible for the GPS system most of us use every day. Space-related programs spawned the technology that enabled the creation of personal computers and smartphones, and much of their use depends on orbiting satellites. Military satellites helped to keep the peace during the Cold War and continue to monitor global security challenges today. Commercial satellites—originally developed for NASA and enhanced by private industry—have provided visual data that has generated vast commercial benefits. Banks depend

Japanese astronaut Akihiko Hoshide after a blood draw, supporting immune system research aboard the ISS. Image credit: NASA

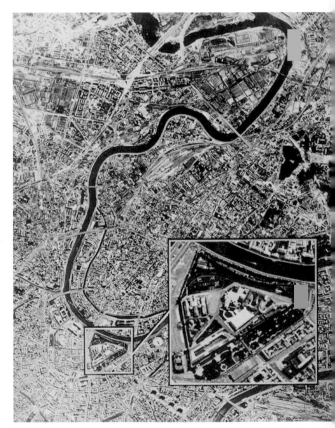

Declassified US military satellite image of Moscow with an insert of the Kremlin from 1970. Image credit: US Air Force

on satellites to carry data across the country and around the globe, speeding and securing the transactions we take for granted. Even our use of the internet requires time tracking provided by satellites. In the near future, global internet access will depend on large constellations of broadband satellites.

Weather tracking and modeling deserves further mention. Working with the Department of Commerce, NASA satellites have provided orbital observations that have enabled the understanding of climate change. Satellites track weather patterns and measure the effectiveness of farming methods and innovations in agriculture, which have helped to feed millions more than we could have dreamed possible before the space age. The collection and distribution of fresh water, rapidly becoming one of this century's most valuable resources, is monitored globally. Aquaculture data—tracking stocks of fish and modeling how to best utilize and maintain a healthy supply of this important food source—is generated from space-based observations. The identification of mineral resources is increasingly performed via satellite imagery. Satellites are also a part of the global traffic control system for aircraft. Even ground transport—trucks, trains, and the like—is monitored from orbit. In short, much of what drives modern civilization is affected, and in most cases improved, by the use of data obtained from

Hurricane Isabel as photographed from orbit in 2008. Most weather tracking and prediction now takes place from space-based assets. Image credit: NASA

satellites. Almost every branch of the government now uses satellites to do its job.

There are also immediate returns to the US workforce via NASA spending, along with other government investment in space activities. The vast majority of NASA expenditures—a higher percentage than any other endeavor except national defense—is spent on American workers and with American businesses. From the manufacture of new space station components; to well-paid high-tech jobs at NASA, its contractors, and cooperating universities; to the wages of the barista at the Starbucks down the street from the Kennedy Space Center, NASA's spending is pumped directly back into the national economy.[14] And while NASA has always hired American aerospace contractors

Lori Garver, former deputy administrator of NASA. Image credit: NASA

an insider's view of NASA spending. She drew a comparison with the initial development of civilian aviation. The US federal government invested heavily in aviation starting in the early twentieth century, and the impact has been huge.

"We now have a trillion-and-a-half-dollar industry that has changed the world," Garver says. "We have billions of people traveling [by air] every year. This was a result of government investment in aviation, and that is what I want to see in space."[15]

Garver's view is that intelligent government investment in space—a balance of the direct funding of technology and research, as well as in the private sector—will offer vast rewards in the long term, just as investments in civil aviation did.

These expenditures in spaceflight and technology provide another far-reaching benefit: trained minds. Every mission sent into space, whether crewed or robotic, has multitudes of skilled people behind it. Many of them are highly educated scientists, engineers, and technicians whose life goals were formed by government space programs. Robert Zubrin, an entrepreneur, aerospace engineer, and avid proponent of the exploration of Mars, noted this in the context of the benefits from the Apollo program workforce.

to build its spacecraft, funds are now being directed to innovative fixed-price contracts with high-tech start-ups such as SpaceX, that will allow us to avoid future spending on Russian launches to the ISS. These expenditures are reinvigorating our national economy via traditional contracts and new and inventive government–private sector partnerships.

Lori Garver, the deputy administrator of NASA from 2009 to 2013, has

If you ask people working at NASA facilities such as the Kennedy Space Center (KSC) and the Johnson Space Center

(JSC), Zubrin notes, the majority will tell you that they were inspired by the Apollo and shuttle programs and the robotic exploration of the planets. "We are much richer today because of the Apollo program and the millions of young scientists and engineers that entered the workforce. Many of them got excited about science because of the lunar landing program. And who were the forty-year-old techno-geeks that built Silicon Valley in the 1990s? Many of them were the twelve-year-old 'mad scientists' making rocket fuel in their basements in the 1960s. They are Apollo's children."[16]

Clearly, the Apollo program inspired many people to seek careers in engineering and the sciences. It also provided inspiration to laypeople across the nation, and around the world, like no other event in modern history. Even staunch opponents of the program—whether civil rights campaigners or critics of excessive federal spending—gave grudging but high praise to the lunar landings, despite any misgivings about the undertaking's expense. For a brief few years after the first Apollo astronauts stepped out onto the lunar surface in 1969, people who had access to the popular media were in awe of the accomplishment. That inspiring quality may, in the end, be spaceflight's greatest achievement.

This inspiration takes many forms. It is instructive to listen to what some of the

Technicians working on the James Webb Space Telescope at NASA's Goddard Space Flight Center in Maryland. Many NASA employees will tell you that the exploits of the agency during the space race inspired them to careers in aerospace and science. Image credit: NASA

people who have traveled to space have to say. Their viewpoints are unique—recall that just over five hundred people have made this trip, after all—and are often formed by seeing Earth from space, a vantage point referred to as the "overview effect." This phenomenon was identified by Stanley Rosen in 1976 and written about extensively by American author Frank White in 1987.[17] The resulting impulse prompts the astronauts to share their feelings about the fragility of the earth, the pettiness of human conflict, and the need to work cooperatively for the betterment of all humanity.

It's not always easy for astronauts to explain their feelings on such matters—

Earthrise from Apollo 8, the image that stirred a global consciousness and led to the identification of the "overview effect." Image credit: NASA

these are people who, quite often, have spent a large part of their lives indoctrinated in the military precision of training as a test pilot (or military aviator), and the words don't always come easily. But they often do speak on the subject, especially later in life. Edgar Mitchell, who flew to the moon on the *Apollo 14* mission, had this to say, thinking back on his perspective from 240,000 miles away from Earth in 1971:

"You develop an instant global consciousness, a people orientation, an intense dissatisfaction with the state of the world, and a compulsion to do something about it. From out there on the moon, international politics look so petty. You want to grab a politician by the scruff of the neck and drag him a quarter of a million miles out and say, 'Look at that, you son of a bitch.'"[18]

That's eloquent, in a salty way.

I gained another perspective during a conversation with Franklin Chang-Díaz, a multiethnic shuttle-era astronaut whose life story reads like a description of a modern-day John Steinbeck character: a salt-of-the-earth immigrant who overcomes all the challenges of entering a new society and reaches, quite literally, the greatest of heights. In his case, this advancement came in the form of his career as a NASA astronaut.

"Since my early childhood and adolescence in Costa Rica and Venezuela,

Edgar Mitchell, Apollo 14 astronaut. Image credit: NASA

Franklin Chang-Díaz, veteran of seven space shuttle missions and now heading a company building advanced propulsion systems for deep-space exploration. Image credit: NASA

I have been living the dream of space exploration—first in my imagination, then in reality. I came to America, alone, as an eighteen-year-old immigrant, with fifty dollars in my pocket. My twenty-five years at NASA changed my worldview and philosophy in that now, more than ever, I believe space holds the key to human survival. I also feel that space must be democratized with people of all nations having the opportunity to go."[19]

Back at the radio station, I seem to have addressed Jeff's concerns and we have moved on to other questions about spaceflight. We could, however, continue to discuss the "why space" question for hours. There are other, broader questions to ponder about the nature of the human species and our place in the universe, and how our perspectives on them can be altered by the exploration of space.

Jakob van Zyl is the director of solar system exploration at NASA's Jet Propulsion Laboratory (JPL). His take on space exploration is not unique but rather is somewhat typical of many who labor in the trade. Specific to his work at JPL, which involves sending robots to the most distant reaches of the solar system, van Zyl thinks about space as it relates to our very origins. He feels that humans are explorers by nature, that it's part of our DNA. He also views this exploration as a kind of time machine. "The exploration of the solar system is the ultimate journey back in time," he says. "The outer solar system objects come the closest to pristine composition of the nebula from which we all formed. So if you really want to understand what it was like in the beginning, you must go out there. Scientifically speaking, that is the real reason . . . to find that primordial material that was part of the beginning of it all, from which we believe everything else formed."[20]

A similar sentiment was shared by someone outside the community of engineers, astronauts, and scientists: a former (and very successful) rock-and-roll promoter named Howard Bloom. His PR firm was one of the most powerful in the record industry of the 1980s, representing the likes of Styx, Prince, and Michael Jackson. He is now an active space advocate

and prolific author. His poetic thoughts summed up those of many in the space advocacy community.

"The main fuel for the space program is not liquid oxygen, hydrogen, or kerosene; it is the human spirit. When something thrills the human spirit, there is a horizon we're going to move beyond—one way or the other. And once we get over any new horizon, things open up to us that we can never imagine."[21]

This is an overarching driver for the exploration impulse. Besides more practical reasons, space exploration is about curiosity. It's about the drive to understand the unknown. It's about pride, human achievement, and bettering ourselves.

There is one final rationale for space exploration, an eminently practical one, that is endlessly argued over by very intelligent people but bears close examination: the very survival of humanity.

Scientists, philosophers, and a few billionaire entrepreneurs have increasingly taken up this call in the past decade. Elon Musk thinks there are compelling reasons to use another world, such as Mars, to create a "backup" of our species. We can make it a second home for humanity, against the likely calamity that will one day preclude living on Earth. As he said in 2016, "There are two fundamental paths: one path is we stay on Earth forever, and some eventual extinction event wipes us out . . . the alternative is to become a spacefaring and multi-planetary species."[22] He sees Mars as the best bet for a colony and is building technology that he hopes will one day take humans there.

Mike Griffin, the Under Secretary of Defense for Research and Engineering and a former NASA administrator, expressed a long-term vision similar to Musk's: "If we humans want to survive for hundreds of thousands or millions of years, we must ultimately populate other planets."[23]

The late famed physicist Stephen Hawking has echoed this sentiment more darkly: "We are running out of space, and the only places to go to are other worlds. It is time to explore other solar systems. Spreading out may be the only thing that saves us from ourselves. I am convinced that humans need to leave Earth."[24] Hawking estimated that, at our current rate of growth and consumption, humans have one hundred more years of qual ity existence on Earth before it becomes uninhabitable.

Robert Zubrin supplied the numbers to conclude the argument for human expansion: "We are using up the earth. The population of our planet currently stands at 7.5 billion people," he noted. "By 2050 it is estimated to grow to almost 10 billion. This will continue to compound our drain on natural resources. This includes metal ores, fossil fuels, and even fresh water. At the same time, we are stressing the earth's environment and atmosphere. Regardless

of where you stand on the climate change issue, we are adding millions of tons of carbon and greenhouse gasses to the air that surrounds us and this will have, and very likely has already had, an impact on global temperatures, which dramat-ically affects everything from agriculture to forest growth to the fishing industry. In short, our planet will reach a point where it will be challenged to support the enormous pressures exerted by a growing population and its needs, byproducts, and waste."[25]

Jeff Bezos, the founder of Amazon and the rocket company Blue Origin, is less keen on relocating the human race—he feels that Earth will forever be the best place for humans to live. But he has other ideas about how the earth will be used. "All of our heavy industry will be moved off planet, and Earth will be zoned residential and light industrial," he said in 2016.[26] He elaborated on his vision at the National Space Society's International Space Development Conference in 2018: "This is not something that we can choose to do. This is something we must do. We need to do a better job of communicating how important [space development] is, because this mission is so big. We need the world to support it . . . this is a 'must-do' and not a 'fun to do.'"

Bezos envisions Earth, once it has recovered from industrial activity, as a pristine global park that people can live on in peace and prosperity, supported by the riches of the solar system.

Freeman Dyson, another physicist and a supporter of human expansion, outlined a similar vision in 2001. "Our job is to help life spread out from this planet and make the rest of the universe as beautiful and varied as the earth . . . Dead worlds may be beautiful, just as deserts may be beautiful, but worlds full of life will give birth to a far wider range of beauty."[27]

I'll close with a discussion I had with Mark Hopkins, a Harvard- and Caltech-educated economist and a leader of the National Space Society, about his thoughts on what the next few decades might look like if we embrace this challenge and begin to move human endeavor into space. He focused on the resources that support human civilization.

"The key to survival is to develop space settlement—humans going there, living there, and working there, to utilize the vast resources we find in space for the betterment of all. The vast majority of resources in the solar system lies in space rather than on the Earth. This is true of both materials and energy . . . The sun produces ten trillion times more energy than what is currently used by the human race. If we could bring just a small percentage of that to Earth, we will have all the energy we need, and it's a clean, inexhaustible resource. There is vast potential for increasing the size of the human

economy by tapping into these resources. Half of the human race currently lives at an income which is more than a factor of twenty-five below the American average. Two billion people are so poor they don't even have electricity. Space offers the resources to increase the global standard of living well beyond what it is in the United States today, and to do this in a way that is environmentally benign. Space makes possible a prosperous and hopeful future for all of humanity."[28]

We have heard a wide range of opinions, from a number of well-respected thinkers of our age, about why space exploration is important. The reasons range from technological advancements that can improve our daily lives, to economics, and even the satisfaction of our basic impulse to explore. Some even posit the movement of humanity off Earth as a hedge against a catastrophic, species-ending event on Earth.

With that understanding in hand, let's take a look at how we have historically gone into space and the state of the art for doing so today.

CHAPTER 4

THE FIRST SPACE AGE

We've been sending humans into space since 1961. We've gotten pretty good at it, though it is still expensive and complicated to achieve. The bounty that awaits us from the true opening of this frontier will not be realized until spaceflight—both robotic and human—becomes far less expensive and the machines reusable. As of today, the only "reusable" crewed spacecraft has been the space shuttle, and it required extensive, and costly, refurbishment between flights. (In hindsight, true reusability was simply too much to expect from 1970s technology.) SpaceX, using a simpler design and new technologies, has proven that its rockets can be reused—it has now reflown Falcon 9s a number of times for paying customers—and reusability is key to lowering costs. As Elon Musk likes to point out, flying one-use, expendable rockets is like flying from Tokyo to New York and then throwing away the airplane. Reaping the economic benefits of space will require quick-turnaround reusable spacecraft—launch, recover, inspect, refuel, and fly, just like airliners. This is a major component of Space 2.0.[29]

But to better understand where we are headed, it may help to get an idea of where we started and where we are today, regarding human spaceflight.

The space race began with the launch of the first Earth-orbiting satellite, Sputnik, by the Soviet Union in October 1957. The US, which had been struggling to launch its own satellite for months, was chagrined and stunned. America had perceived itself as the leader in technology since World War II and, by the late 1950s, considered itself the most technologically advanced nation on Earth. To be humbled by the Soviet Union, the only other superpower and an avowed enemy of the West,

was unthinkable. Yet it had happened, and reaction in the press was shaming both internationally and at home.

The US finally launched its own satellite, Explorer 1, just a few months later in January 1958.

More satellites followed from both countries, but the new goal was to put a human in space. Both countries turned to their burgeoning nuclear missile arsenals for a rocket that would be powerful

The Soviet Union launched Sputnik on October 4, 1957, kicking off the first space age. Image credit: NASA

Von Braun explains Explorer 1 prior to launch. Image credit: NASA

The US followed the launch of Sputnik by launching Explorer 1 on January 31, 1958. After the successful launch, a press conference was held at NASA's Jet Propulsion Laboratory (JPL). From left to right: Bill Pickering, a founder of JPL; James Van Allen, famed space scientist; and Wernher von Braun, who would go on to build the Saturn V moon rocket. Image credit: NASA

The Huntsville Times

HUNTSVILLE, ALABAMA, WEDNESDAY, APR. 12, 1961

Man Enters Space

So Close,
et So Far,'
ghs Cape

S. Had Hoped
Own Launch

Soviet Officer
Orbits Globe
In 5-Ton Ship

Maximum Height Reached
Reported As 188 Miles

To Keep Up, U.S.A.
Must Run Like Hell'

Praise Is Heaped
On Major Gagarin

First Man To Enter Space

Newspaper headline announcing the launch
of Yuri Gagarin on April 12, 1961. He was
the first human to orbit the earth. Image
credit: NASA

Yuri Gagarin in 1961. Image credit: NASA

enough to lift a small space capsule with a single passenger. The Russians had more experience with large rockets at that time, and they launched Yuri Gagarin into a single-orbit flight in 1961. The US, stung at once again being shoved into second place, flew Alan Shepard in a suborbital trajectory just weeks later, and then orbited John Glenn in early 1962.

After being beaten into space by the Soviet Union twice, the new president, John F. Kennedy, consulted with his advisors and NASA to identify a space endeavor that the US stood a good chance of completing before the Soviets. This led to his famous 1961 pronouncement that the US would "land a man on the moon and return him safely to the earth," with a completion date of "the end of this decade." The space race was off at full tilt—born not of scientific goals or a passion for exploration, but by a massive geopolitical struggle between rival superpowers. It was a non-shooting war to prove which country had the better system of government and education, and whether a military-dominated program such as the Soviet Union's could best a civilian program like the US's. It was, of course, also a test of the technological prowess of both nations.

The next eleven years saw rapid advancements in spaceflight. The Soviets flew three different, increasingly capable space capsule designs with more improved capabilities. This culminated in the Soyuz,

Alan Shepard prepares to become the first American to fly into space, albeit suborbitally, wedged inside the tiny Mercury capsule. Image credit: NASA

The launch of Mercury Redstone 3 (MR-3), carrying Alan Shepard on a fifteen-minute suborbital flight. This mission put the US into the manned spaceflight business at last. Image credit: NASA

originally created to win the race to the moon. A version of the Soyuz is still in use to this day. The Soviets also built a series of increasingly powerful rockets, which led to their massive N1 booster, again intended to help them beat the US to the moon. That rocket, unfortunately, never flew successfully and was abandoned in the 1970s, after the US had won the race for which the N1 had been designed.

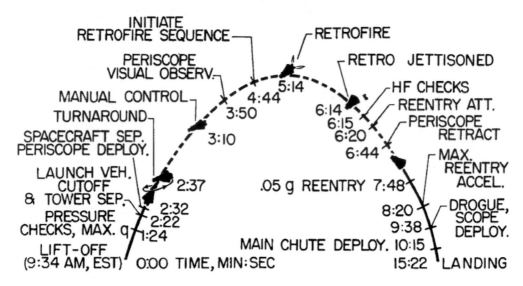

A vintage NASA timeline of the MR-3 flight of Alan Shepard. Image credit: NASA

In the United States, NASA flew the Mercury spacecraft, large enough for a single astronaut, then Gemini, which held two people, through the mid-1960s. After mastering the ability to change orbits, rendezvous, and dock two spacecraft during the Gemini flights, the Apollo spacecraft began operations in 1968. The US also built increasingly larger rockets, culminating in the Saturn V moon rocket, which flew between 1968 and 1973.

The first humans landed on the moon on July 20, 1969, and depending on your definition, the space race drew to a close when the last Apollo mission returned to Earth from the moon in 1972. By the mid-1970s, both the Soviet Union and US had orbiting space stations and were

continuing to explore the planets with robotic spacecraft.

In the 1980s, America's space shuttle, intended to be a less expensive alternative to the Apollo spacecraft and Saturn V rocket, became operational. The Soviet Union continued flying the proven Soyuz and built a modular space station called Mir. Both countries carried out increasingly ambitious research projects in Earth orbit, and both were focused on human factors in extended spaceflight: How did the human body adapt to long-term weightlessness? Why did so many astronauts and cosmonauts get sick, especially with vertigo, after reaching orbit? What were the roles of the inner ear, vision, and other senses—all cultivated during an astronaut's life in a one-g Earth

Alexei Leonov becomes the first human to walk in space on March 18, 1965. While successful, it almost cost him his life when his space suit ballooned up and he was barely able to reenter the capsule. Image credit: NASA

Gemini 7 in orbit as seen from Gemini 6, during the first rendezvous in space in December 1965. Image credit: NASA

A Saturn V rocket the night before launch. The streak to the left is lightning. Image credit: NASA

Possibly the most memorable image from the Apollo 11 mission. Neil Armstrong is reflected in Buzz Aldrin's visor during their moonwalk. Image credit: NASA

A mockup of Viking 1 in a test bay on Earth. Viking was the first spacecraft to successfully land and operate on Mars. Image credit: NASA

The US space shuttle docked to the Russian Mir space station in 1996. The shuttle flew to Mir ten times, an early example of international cooperation in space. Image credit: NASA

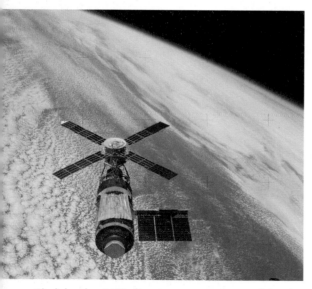

Skylab, the US's first space station, orbited in 1973. It was visited by three crews for increasingly long durations, then allowed to reenter Earth's atmosphere in 1979. Image credit: NASA

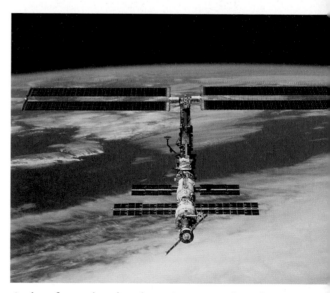

A shot from the shuttle as it approaches the first phase of the International Space Station in 2000. Image credit: NASA

The combined crews of the shuttle and Mir get together for a space selfie. Image credit: NASA

A Soyuz spacecraft docked to the International Space Station in 2015.
Image credit: NASA

environment—in space sickness? These and other medical and flight-adaptation questions were studied during orbital missions of ever-lengthening duration.

After the fall of the Soviet Union in 1991, the shuttle began a series of missions to deliver US astronauts to the Russian Mir space station. This program, which kicked off in 1995, demonstrated a new standard of international cooperation between the US and the new Russian Federation and continued through 1998.

By 2000, the first crew had arrived at the new ISS, the result of an unprecedented collaboration between the US, Russia, the European Space Agency, Japan, and Canada. Crews were delivered by both the Soyuz and the shuttle. The shuttle continued to supply and expand the station until its final flight in 2011.

Since then, only the Soyuz has been capable of ferrying passengers to the ISS.

This brings us to the present. Russia is designing a replacement for Soyuz, and the ISS is scheduled to operate until about 2025. The actual date of decommissioning depends on both political and technical factors, currently under consideration by NASA and its international partners. China flew its first humans into space in 2003, using its own updated version of the Soyuz spacecraft, which it licensed from Russia. China has also flown two small experimental space stations.

Let's summarize the current international spacefaring fleet of human-rated spacecraft. A reminder that the US does not have any current operational spacecraft to transport people, at the time of publication.

HUMAN-RATED SPACECRAFT
CURRENTLY FLYING

SOYUZ

The venerable Soyuz spacecraft has been in service since 1967 and has proved itself a reliable way to reach orbit since then. It is now in its fourth generation and carries a crew of three. The Russian space agency Roscosmos is working on a more modern and spacious replacement, but when this new model will begin flying is not entirely clear. The current Soyuz launches on a booster of the same name. Neither the capsule nor the rocket are reusable.

A Soyuz spacecraft carrying a crew of three, closing in on the International Space Station in 2005. Image credit: NASA

SHENZHOU

Shenzhou is an updated and larger version of the Soyuz capsule, built and flown by the Chinese space program. It also carries three people and rides on a Chinese booster called the Long March 2. Neither the booster nor the capsule are reusable.

The Shenzhou spacecraft, as displayed in the Shanghai Science and Technology Museum. Image credit: Rod Pyle

HUMAN-RATED SPACECRAFT EXPECTED TO BE FLYING BY 2020

DRAGON

SpaceX has been delivering cargo to the ISS with its robotic Dragon spacecraft since 2012. An improved version of the capsule, the Dragon 2, is scheduled to begin delivering crews of four astronauts to the ISS in 2019 or 2020, and it's designed to carry up to seven. The Dragon 2 is intended to be fully reusable for at least ten flights and will launch on the Falcon 9, which is also reusable. The cargo-only Dragon currently flies under contract to NASA.

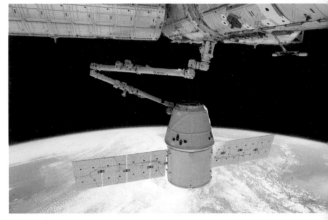

A SpaceX Dragon capsule is seen delivering cargo to the International Space Station under contract with NASA. Image credit: NASA

STARLINER

Boeing's CST-100 Starliner is, like the competing Dragon 2, intended to provide affordable transport to and from the ISS for crews of four astronauts and can also carry a maximum of seven. It will initially launch on the Atlas V rocket built by United Launch Alliance, a partnership of Boeing and Lockheed Martin, but it's intended to be compatible with other boosters as well. It is designed to be reused up to ten times and will also fly under contract to NASA. Its primary booster, the Atlas V, is not reusable.

Boeing's Starliner capsule is scheduled to begin service to the International Space Station in 2019 or 2020. Image credit: NASA

NEW SHEPARD

Blue Origin, Jeff Bezos's rocket company, has made eight flights of its New Shepard spacecraft, all suborbital. Like the Soyuz, both the capsule and rocket carry the same name. The New Shepard is still in testing, but Blue Origin has announced its intention to carry six passengers on short tourist flights by the end of 2019. Larger orbital rockets, the New Glenn and massive New Armstrong, are also being planned. The rockets and capsules will be reusable.

UNITY

Virgin Galactic, Richard Branson's space tourism start-up, is test flying its suborbital rocketplane and plans to be flying tourists by 2019 or 2020. The rocketplane is capable of carrying multiple passengers, is fully reusable, and is dropped from a carrier plane at altitude.

Blue Origin's New Shepard suborbital rocket has flown nine times with eight successes and should be ready for tourist flights by 2019. The spacecraft system is fully reusable. Image credit: Blue Origin

HUMAN-RATED SPACECRAFT EXPECTED TO BE FLYING BY 2025

DREAM CHASER

Sierra Nevada Corporation's Dream Chaser spacecraft was the third commercial entry into NASA's competition for the delivery of crew to the ISS, but it was unsuccessful in the first round of trials. NASA felt it was not yet sufficiently mature technology.[30] NASA has subsequently suggested that it will use the spacecraft for uncrewed cargo runs to the ISS and will reevaluate the craft for human crews in the future. This is not a capsule, but rather a "lifting body" design—it has stubby wings but derives most of its lift from the shape of its hull. The Dream Chaser will glide to a landing just as the shuttle did. It is reusable and will be capable of launching on a variety of rockets.

The Dream Chaser has had a long developmental path, but is currently in flight-testing and should be flying cargo to the International Space Station in the next few years. Image credit: NASA/Ken Ulbrich

SLS/ORION

NASA's Space Launch System and Orion deep-space capsule completes the list of near-term human spaceflight options. Orion was test flown without a crew in 2014 on a Delta IV rocket. It is the largest of the human-rated spacecraft discussed here. Orion is designed for extended missions to the moon and possibly Mars. The SLS booster is planned to be versatile enough to carry Orion and heavier payloads to orbit and beyond, with its final iteration capable of lifting more mass than the Saturn V of the 1960s. While both SpaceX and Blue Origin also have large rockets planned (and, in SpaceX's case, currently under construction), the SLS is the largest rocket in NASA's existing plans. The SLS rocket is not reusable, but the Orion is intended to be.

NASA's Orion spacecraft will fly with its first crew in the early 2020s. Orion is designed for deep-space exploration as well as Earth-orbital use. Image credit: NASA

The early version of NASA's Space Launch System, or SLS, is called Block 1. The later Block 2 will have increased payload capacity. SLS is scheduled to begin flying with crews in the early 2020s. Image credit: NASA

SPACE STATIONS

INTERNATIONAL SPACE STATION

Started in 1998 and completed in 2011, the ISS is the largest, most complex, and most expensive space project ever undertaken. It has 33,000 cubic feet of interior space and has been hosting crews of up to six continuously since 2000. It is expected to continue operations until 2025.

The International Space Station is the most expensive object ever built. Permanent occupation began in 2000. Image credit: NASA

TIANGONG

The Chinese space program has launched two space stations so far, with a third planned for the early 2020s. Tiangong-1 was a small experimental station visited by Chinese crews on three flights, but it lost contact with ground controllers in 2016, reentering the atmosphere in 2018. Its successor, Tiangong-2, is in orbit, has hosted successive crews, and has tested robotic resupply capability. Both were primarily engineering test missions but have also housed life-science studies. Tiangong-1 and -2 were single-module stations; the new Chinese large modular space station, scheduled for about 2022, will provide a larger, multimodule orbital outpost.

MIR 2

The Russian space agency has been discussing a new orbiting station for decades—in fact, some of the Russian components of the International Space Station were originally intended for Mir 2, which was put on indefinite hold after the fall of the Soviet Union. When the ISS is decommissioned, Russia may detach some of its hardware for repurposing in orbit as part of a new project of its own design.

BIGELOW AEROSPACE

Bigelow Aerospace, founded by multimillionaire Robert Bigelow, has been developing expandable orbital modules for many years. These launch in a flattened configuraton and are expanded—essentially inflated—once they reach orbit. Originally designed for orbital tourism, two subscale, uncrewed Bigelow modules were flown in 2006 and 2007 by Russian rockets and tested in orbit with some success. A third unit, the Bigelow Expandable Activity Module (BEAM), was docked with the ISS in 2016 and is currently undergoing long-term testing in space. The near-term goals of the company include the deployment of the B-330, a 330-cubic-meter module that they hope will become part of a lineup of standardized crew habitation units for use in orbit and beyond. Bigelow proposed a partnership with NASA in late 2017 that would send two of its B-330s into orbit around the moon in the 2020s.

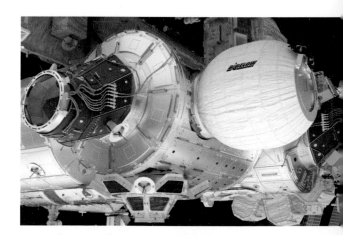

A rendering of Bigelow's BEAM module, currently being tested on the International Space Station. Image credit: NASA

So we have a magnificent, if aging, International Space Station, and a much smaller, experimental Chinese orbital habitat currently in space, along with a small Bigelow test module attached to the ISS. Two systems—one Russian, the other Chinese—for flying humans into space are operational, and various up-and-coming private suppliers in the US are preparing their own systems. Russia and the US handle the ISS's cargo needs robotically. NASA's SLS and Orion should be flying crews into space in the early 2020s. Bigelow, a company called Axiom Space, and other entities are investigating the lofting of privately financed research and tourism platforms into orbit.

Regarding rockets, the US has the Atlas V and Delta rockets from ULA, and SpaceX's Falcon 9 and Falcon Heavy. SpaceX and Blue Origin are both developing larger rockets (and, in SpaceX's case, spacecraft) for the 2020s. The Russians, Chinese, Japanese, and Indians each have their own boosters. These rockets will be covered in more detail in subsequent chapters.

The above list outlines the current state of human-rated spacecraft and their immediate futures. What can we expect Space 2.0 to add to the mix in the next few years? The Starliner and Dream Chaser are designed specifically for operating in low Earth orbit. SpaceX's Dragon 2 capsule is intended to be useful out to lunar orbit.

NASA's Orion and SLS should be capable of a diverse set of mission profiles, including journeys to Earth orbit, the moon, Mars, and near Earth asteroids.

In the past, spacecraft have been very destination driven. The early Soviet spacecraft, Vostok and Voskhod, and NASA's Mercury and Gemini capsules were designed for low Earth orbit. Soyuz and Apollo were intended for the moon. All were designed to be perfectly suited to their specific goals, though, in practice, the Soyuz ended up being very adaptable. As we move ahead, new spacecraft designs will be increasingly flexible regarding uses and destinations. This is another key to achieving affordable spaceflight. Nevertheless, what we build and fly will depend to a large degree on where we intend to go and what we will do once we are there.

Now that we understand more about the current NASA, commercial, and international human spacecraft inventory, let's talk about some of the preferred destinations in space. Where do we want to go and what makes those destinations appealing?

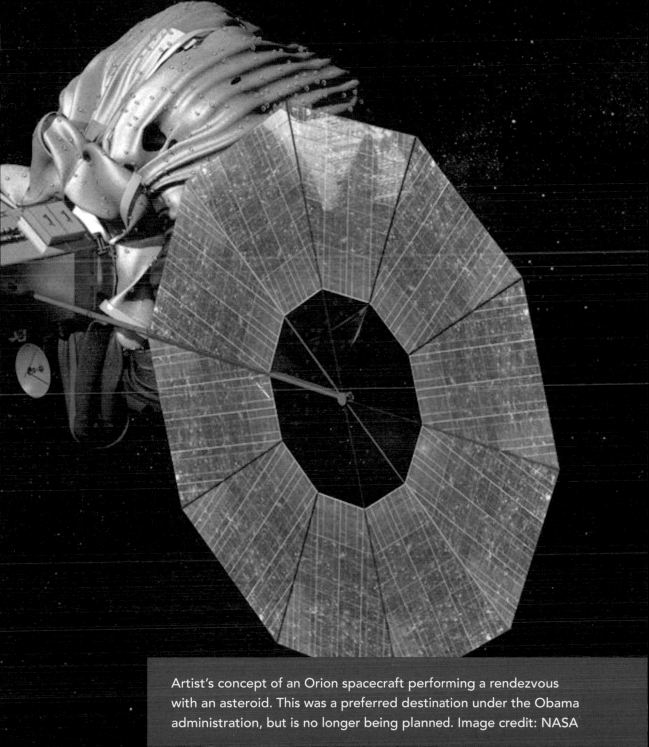

Artist's concept of an Orion spacecraft performing a rendezvous with an asteroid. This was a preferred destination under the Obama administration, but is no longer being planned. Image credit: NASA

Image credit: James Vaughan

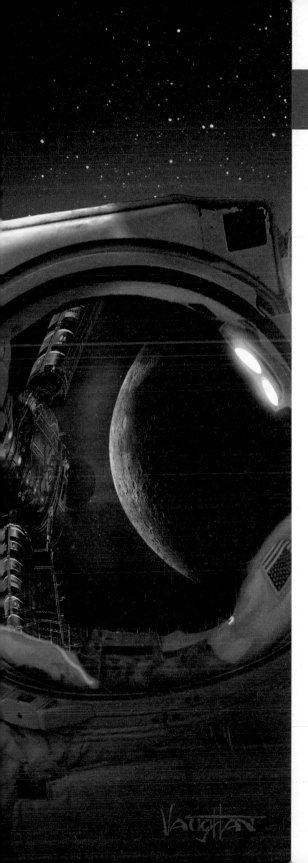

CHAPTER 5

DESTINATIONS

In terms of human spaceflight, there are eleven general destinations that make sense, given the technology that is likely to be available in the next twenty to thirty years.

- Earth Suborbital: A trajectory followed by a spacecraft that does not reach orbit. The early Mercury missions and the X-15 rocketplane flights were suborbital. In the immediate future, suborbital trajectories will be primarily useful for space tourists. Virgin Galactic's and Blue Origin's tourist flights will be suborbital.

- Low Earth Orbit (LEO): Orbits that range from 100 to 1,200 miles above Earth's surface. This is where the ISS and Soyuz spacecraft operate. LEO is also where many satellites, both scientific and commercial, operate.

- Geosynchronous and Geostationary Earth Orbit: Orbits about 25,500 miles from Earth, where a satellite orbits at the same period the Earth turns, causing it to either remain stationary relative to the earth's surface (geostationary) or reappear in the same place every twenty-four hours (geosynchronous). This is not considered a prime destination for human spaceflight but is very useful for communications, weather monitoring, and military satellites.

- Cislunar Space: Anywhere between geosynchronous Earth orbit and slightly beyond the orbit of the moon. This technically includes lunar orbit and some Lagrange points. Both of these are detailed below for clarity.
- Lunar Orbit: An orbit around the moon at any altitude.
- Lunar Surface: Surface operations on the moon.
- Lagrange Points: Stable points in space where spacecraft can retain their relative position to two bodies, such as Earth and the sun or Earth and the moon, with very little correction to their trajectory.[31]
- Asteroids: Orbits near, or landings on, asteroids.
- Mars Orbit: Similar to LEO, a useful low Mars orbit (LMO)

NASA's Lunar Orbiting Platform-Gateway is an example of a program in cislunar space, orbiting the moon at a high altitude. Image credit: NASA

could be about 180 miles. Mars also has a geosynchronous orbit; due to its lower mass than Earth, this is about 8,000 miles above the surface.

- Martian Moons: Orbits around, or landings on, the moons of Mars, Phobos and Deimos.
- Mars Surface: Operations on the surface of Mars.

There are compelling arguments for the utility of any of these destinations, but this gives you a general idea of places that make sense for crewed journeys beyond Earth orbit in the next few decades.

Predictably, people who plan space missions—from NASA, to international space agencies, to academicians and even well-informed private citizens—have strong opinions about which of these destinations make sense and will be most useful. Any of the regions on the list have pros and cons. With limited funds, it is likely that NASA, or a consortium of NASA and other international space agencies, will probably only undertake human spaceflight missions to one, two, or at most three of these regions between now and 2050. Although robots have already gone to many of these places (and well beyond), and are likely to visit them all again within this time span, it is far more expensive to send humans, so difficult choices must be made. Private-commercial space entrepreneurs such as Elon Musk can attempt to go wherever they choose, and Musk has identified human voyages to Mars as a priority for SpaceX, but how quickly they will accomplish this remains to be seen.

Human destinations in space are a subject of hot debate. Low Earth orbit will continue to be the most heavily trafficked region regardless of where we go in deep space, since it is the most affordable and technically the easiest to achieve. Lots of productive work can be performed there, especially by commercial, research, and military satellites. China and Russia are both planning new crewed scientific

View of an aurora from low Earth orbit, taken from the International Space Station. Image credit: NASA

View of Hayn Crater near the lunar north pole, a likely location of water ice. Image credit: NASA

outposts in LEO, and various private enterprises, primarily in the US, are planning both scientific and tourist-oriented orbital stations. So low Earth orbit is a given. There will be a vast amount of activity there, both human and robotic, indefinitely.

Beyond low Earth orbit, the most favored candidates for human spaceflight are the moon and cislunar space, and Mars and its orbital regions. There are factions within the larger spaceflight community that favor one or the other, and they tend to be passionate about their selection. For example, there is a large "Moon First" crowd, including the current president's administration, that is convinced that returning humans to the moon, only 240,000 miles distant, is our best next move. We can reach that body within a few days, instead of the six or seven months it would take to get to Mars. We understand the surface of the moon, having extensively examined rocks and soil brought back from its surface. In the event of an emergency, Earth is relatively near, and that proximity allows for a quick return.

There are other advantages, too. We know how to operate on the lunar surface with both humans and robots. We know there are metals and glass to be refined from lunar ore, and we are fairly certain that we can extract water from ice near the lunar poles. Water is particularly valuable for humans, as it is useful for life support, consumption, and providing breathable oxygen. That water can also be used to make rocket fuel once it is separated into hydrogen and oxygen, which is a familiar process. There are mountain peaks on both poles that receive sunlight up to 90 percent of the time, which are good locations for solar power stations. The Moon First crowd also contends that the moon is the perfect place to test and evolve the technologies needed to move out to Mars and, ultimately, beyond.

Not so fast, say the Mars First advocates. If we spend our money and resources researching and developing bases in orbit around, or on the surface of, the moon, it could be many more decades before we reach Mars—there is, after all, only so much money for space programs. Mars has a thin atmosphere that will help to mitigate radiation. Mars also contains lots of resources useful for human travelers and colonists. Water will be extractable from ice below the surface of Mars as well as its atmosphere, and with water you can make oxygen, useful for life support, and hydrogen, for rocket fuel. Mars has higher gravity than the moon, 38 percent of Earth's as opposed to about 16.7 percent, which may be healthier for humans.

There's one more reason, albeit a more subjective one, some would prefer we go to Mars first: It's a planet. Mars contains about the same dry land area as Earth, with a far greater range of surface topography than the moon. And Mars was once much more Earthlike than it is now (the moon has always been dry and airless), and so it may have much to teach us about planetary evolution and decline. The list topper is, of course, the possibility that Mars may have hosted simple forms of life in the past (and may still).

There is yet another group that favors orbits around other worlds as opposed to landings. Orbits, whether around the moon or Mars, are much easier to accomplish than landings. Without expensive and complex landers required, there is also no need to launch all the extra fuel needed for them to return home; recent estimates of the amount of fuel required for a crewed vehicle to leave the surface of Mars are about 33 tons.[32] The planet or moon being circled can be extensively studied from on high, especially if robots

Mars is favored by many for human settlement. Image credit: NASA/Pat Rawlings

A Mars orbiting station proposed by Lockheed Martin, the Mars Base Camp. Image credit: Lockheed Martin

on the surface are controlled from orbit. Ultimately, orbiting stations will probably be equipped with centrifuges that will create artificial gravity, making it easier to keep the crew healthy. Returning to Earth is far easier from a lunar or planetary orbit as well. Finally, these "orbiting waystations" can ultimately serve as points of departure for surface operations on the body they orbit.[33]

Similar arguments are made for crewed stations at some of the Lagrange points—specifically L1 and L2. Both are reasonably stable places to put a space station (only occasional position adjustments would be required), with L1 between the earth and the moon and L2 on the far side of lunar orbit. L1 has permanent sunshine, perfect for generating power and facilitating solar astronomy. Similar work to that performed on the ISS, such as measuring the effects of long-term exposure to microgravity, can be performed at L1 or L2. Both may also prove to be useful places to store fuel to enable travel deeper into the solar system, as well as staging points for crews traveling to and from the moon and Mars. They are also thought to be useful places to move asteroids for mining and possibly for the storage of resources. L4 and L5, respectively leading and trailing the moon along its orbit, are regions where small asteroids are thought to gather. For these reasons, some feel that these and other Lagrange points

Robotic probes reconnoiter an asteroid for possible ore of value. Image credit: NASA

are also good places to use as waypoints for deep-space colonies.[34]

Finally, there are the asteroids themselves. There is much scientific knowledge to be gained from studying them, since asteroids represent fragments from the early solar system that are unblemished by exposure to weathering on a planet's surface. Many of them contain metallic ores that can be mined and made into tools and parts for spacecraft. Some contain rare earth elements, valuable for manufacturing semiconducting electronics, that are increasingly difficult to find on Earth. Asteroids also hold water ice that can be processed in the same way as ice on the moon and Mars, providing handy sources of fuel and life support. If placed at a Lagrange point, an asteroid can be a productive place to position an outpost.

What these destinations have in common is that they are all difficult to reach. We know how to get to these places—the basic technologies needed to reach most of them are well understood. But sending people to these locations for anything longer than a short expedition is harder. Those of you who are old enough to recall the Apollo program, during which expeditions to the moon were leaving Earth as frequently as every two months, may also recall plans made at the time to reach Mars within a decade and build bases on the moon and Mars by the end of the twentieth century. These plans, of course, did not come to pass.

Still, all these destinations, and the programs that will enable us to reach them with manned flights, are being actively discussed inside NASA, within governments and universities, by space entrepreneurs, and in international space agencies outside the US. On the NASA side, taking into consideration the recent policy statements from the National Space Council, it appears that parallel programs are likely to emerge. As the ISS prepares to wind down operations in 2025, NASA will work with private contractors to build and operate outposts in lunar orbit and eventually on the lunar surface. These will initially be scientific in nature, with a gradual evolution of commercial activities. NASA will also continue to pursue both robotic and human exploration

deeper into the solar system, and it plans to eventually send humans to Mars.

Multiple goals will likely be pursued by different nations and private groups as well. China, for example, has stated that it intends to land humans on the moon between 2030 and 2040. The European Space Agency wants to create an international "Moon Village," with US, Russian, and possibly Chinese cooperation, by the 2030s. The Russians still talk about a crewed Mars landing as a goal, as does NASA. How much international cooperation can be mustered may depend upon national pride and political realities that exist well outside the control of the various space agencies.

Another big variable is what entrepreneurs like Elon Musk will seek to accomplish with regard to human missions to the moon and Mars. Currently, his primary goal is to send humans there by the mid to late 2020s, which could theoretically beat NASA to the Red Planet.

The next five years should be interesting ones for spaceflight.

We now find ourselves almost as many years from the age of Apollo as the first moon landing was from the initial flights of the Wright brothers. Clearly, large-scale, long-term commitment to goals in space is a difficult thing to maintain. Making these the necessary choices and then committing to accomplishing the goals we set will be hard, but decisions must be reached so that we can move forward efficiently and safely. The decision makers will, ideally, pick a destination—or destinations—where we can both remain and prosper. Space, whether in Earth orbit, in the emptiness between worlds, or on the surfaces of distant planets or moons, is a hostile environment. Surviving, and thriving, is critical.

Expedition-style journeys to the moon and Mars are likely to precede more permanent settlements. Image credit: James Vaughan

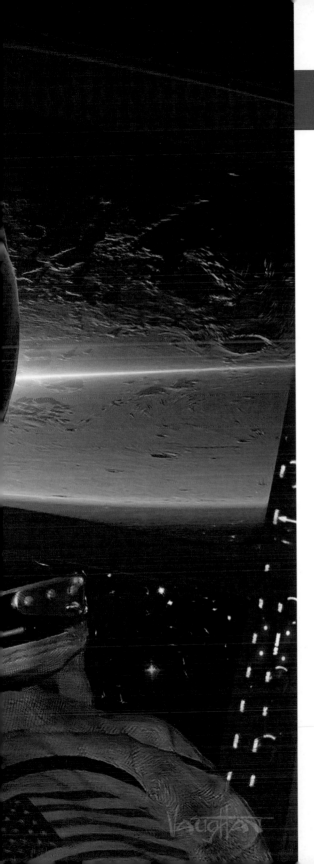

CHAPTER 6

THE HUMAN FACTOR

During the early space race, the US and Soviet Union were competing to get a human into space as quickly as possible. To loft their capsules, they modified their nuclear missiles to carry heavier loads. The spacecraft themselves were clean-sheet, or new, designs, and were in essence just large enough to protect a single man during the journey into orbit. Besides concerns about launch and reentry, the primary challenge was to keep that person alive. Unlike the repurposing of missiles for launching people, this was a new area of technological development—the closest terrestrial analogies were submarines and high-altitude aircraft. Of the two, high-altitude flight came the closest to the demands of spaceflight.

Pressure suits were one challenge. Both countries repurposed designs for pressure suits that had been developed for pilots of high-altitude aircraft, and the resulting garments were sufficient for short orbital spaceflight. In most cases, these suits were intended only to protect the pilot for brief periods of vacuum exposure, in case of an emergency. They were designed to be worn

69

Pete Knight beside the high-altitude X-15 rocketplane, mid-1960s. At this time, pressure suits were designed to keep the pilot alive in low-pressure envinronments—they were, essentially, just body-shaped balloons. Image credit: NASA/USAF

for a matter of hours at most and would only be pressurized when necessary. The goal was protection of the pilot, not mobility or comfort.

As both space programs evolved, flights became longer, lasting days and, later, weeks, and other variables needed to be considered. Besides protecting the pilot

inside the spacecraft from a sudden drop in pressure, the elimination of waste while wearing the suit became an issue. The original designs were basically sealed bags with a front or rear zipper. Diaper-like garments were added, along with methods for urinating inside of, and later through an opening in, the suit. Still later, elimination of solid waste was accomplished with plastic bags, an unwieldy and awkward procedure inside tiny space capsules—not to mention the odor.

As spacewalking or "Extra Vehicular Activities" (EVAs) became a part of spaceflight, the functions of space suits expanded again. In addition to maintaining pressure in the event of cabin pressure loss, the suit now needed to protect the wearer from the harsh extremes of temperatures outside the spacecraft—about 250°F on the sunlit side, −250°F (or colder) on the shaded side. These suits also had to be flexible enough to allow the wearer to perform work while it was pressurized, which turned out to be very difficult—much more difficult than was expected by designers. It's very complicated to make a garment that can be sealed, pumped up to about five pounds per square inch, and remain flexible. All manner of clever designs were prototyped and tested, but few functioned as hoped. The prototypes, once inflated, were rock hard and tended to spring back into a position that made the wearer look like a

Ed White during America's first space walk, from Gemini 4 in 1965. Prior to such activities, space suits were primarily protection against spacecraft depressurization. With the advent of spacewalking, the suits needed to be stronger and more flexible. Image credit: NASA

humans from immediate exposure to the extremes of space. But space suits are just one part of surviving outside the protection of Earth. As programs advanced beyond the race to land on the moon, and on to building space stations, humans would be living and working for months in weightless conditions. This presented new challenges.

By the early 1970s, long-term Earth-orbiting space stations took center stage. The Soviet Union launched its first orbiting station in 1971, called Salyut. The US followed with the larger Skylab in 1973. These missions revealed, for the first time,

John Young jumps while saluting the flag during the mission of Apollo 16 in 1972. The lunar EVA suits were rugged and reasonably flexible, a huge advance over their Gemini predecessors. Image credit: NASA

gingerbread man. All manner of bellows at the elbows and knees were tried; even cable and pulley mechanisms were used to help pull the arms into a workable position.

Nonetheless, by the time astronauts were walking on the lunar surface in 1969, most of these problems had been addressed. We knew how to protect

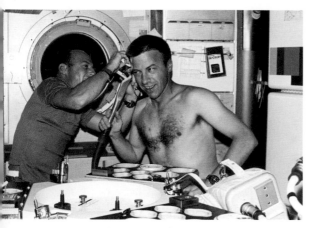

Pete Conrad trims Paul Weitz's hair during the Skylab mission in 1973. Weitz is holding a vacuum device to catch the trimmings. The crew of three stayed in space for twenty-eight days, at that time, the longest US mission to date. Image credit: NASA

the effects of long-term weightlessness on the human body.

By 1975, crews had spent months in space aboard Salyut and Skylab. The longest flight consisted of four grueling months aboard Salyut 6, which had only about 3,200 cubic feet of cramped interior space, or about a third more volume than a standard cargo container. Both countries kept meticulous records of the various effects of weightlessness on the human body, though record-sharing between the superpowers was minimal due to the ongoing Cold War.

NASA flew three crews to Skylab, each staying longer than its predecessor. The medical experiments were the first

Normal Bone Density   Low Bone Density

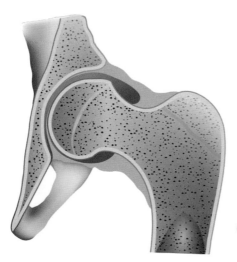
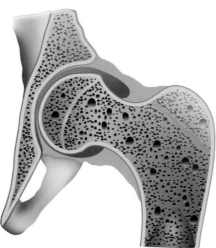

Cutaway view of the results of bone-density loss, which can result in brittle bones. Image credit: NASA

Scott Kelly after months in space. His face is far puffier than normal due to the redistribution of body fluids while weightless. They collected in his midsection and his head. Image credit: NASA

extensive US tests performed on bone loss, reduction in muscle mass, and other changes in the human body when exposed to long-duration weightlessness. The Soviet Union performed similar experiments and recorded similar results.

By the time the Soviet Union fell in 1991, its Mir space station had taken the lead in continuous spaceflight. It had been in orbit for five years, and medical records were now more freely shared with the US. It had become increasingly clear that prolonged weightlessness was not kind to humans. Our bodies, used to spending virtually all their time in a one-gravity environment, don't behave properly in zero-g.

But the resulting bloating and puffiness are a mere inconvenience. More worrisome were the changes in bone density that occurred during long-term weightlessness. As mentioned in Chapter 2, the human body, when not confronted with the task of standing upright on Earth, sheds calcium, leaching it from bones and eliminating it through urine. This weakens bones over the long haul. Add to this the withering of muscles, which no longer have to work to overcome the effects of a one-g environment, and you can have real problems upon your return to Earth. These and other issues began to cause concern to flight surgeons—the astronauts' doctors. Let's look at some possible solutions to these issues through the eyes of a NASA medical professional.

Bill Tarver is a NASA flight surgeon based at the Johnson Space Center in Houston. He graduated with an MD from the University of Texas, then spent nine years in the Air Force. After this, he entered private practice, but, as he put it, "I had a couple of acquaintances at the Johnson Space Center who pestered me to come and visit them. And after five years of private practice, doing occupational medicine—which is a little more mundane—I went and visited JSC and I saw a picture from the moon showing the Earthrise that was signed by Neil Armstrong, Buzz Aldrin, and Mike Collins. I said, 'You mean I could meet Neil

Armstrong?' and they said yes. Well, that just lit my fire, and here I am."[35]

As a part of this elite cadre of about twenty NASA space doctors, and as the former president of the Space Medicine Association, he is in a unique position to observe and evaluate the effects of extended weightlessness on the human body, as well as the effects of recirculated air, a restricted diet, and levels of radiation that, while lower than those experienced in deep space, are still much higher than we receive on Earth. His working days are spent in the medical facilities at JSC, taking care of active and former astronauts. This allows the doctors to monitor the lingering effects of spaceflight.

"When you're floating around in microgravity on the space station, things happen that you wouldn't imagine," Tarver says. "For example, a big bubble of CO_2 gas can collect in front of your face if there is not adequate air circulation. So if you're behind a panel on the space station, and if you don't have adequate ventilation moving the air around, you'll create a CO_2 bubble right there that can cause health problems."

A number of such potential hazards have been discovered since people began spending longer periods in orbit. Some are

There is a lot of instrumentation aboard the ISS and plenty of places for "bubbles" of exhaled carbon dioxide to collect if an astronaut works in one place for too long, so proper air circulation is critical. Image credit: NASA

minor, while others more serious, but the issue of overall physical degeneration is the primary concern. "When it comes to floating in space, it's an absolutely unique condition that you can't reproduce on Earth . . . [in space] there's nothing pulling on your blood vessels or on your body fluids and we can't really simulate that on Earth," Tarver says. This is why detailed studies of people aboard the ISS continue to be so valuable.

The most recent and comprehensive study of the long-term effects of zero-g occurred in 2015, when US astronaut Scott Kelly spent almost a year aboard the space station. While Russian cosmonauts had flown slightly longer durations in the past, Kelly's mission was unique in that he had an identical twin brother, Mark Kelly, who had also been an astronaut and agreed to be monitored by medical specialists while his brother was in space. This opportunity to have a control subject on Earth—right down to identical DNA—while the experimental subject was in space, along with extensive medical history on both men, was a first in space physiology studies. As it turned out, a number of surprises were in store for the NASA doctors.

Scott Kelly experienced changes to his eyesight, which in his case appear to be permanent (women appear to experience this less than men, for reasons that are not yet clear). He now wears reading glasses due to his eyeballs having changed shape

Scott Kelly (right) and twin brother Mark Kelly (left), both NASA astronauts, were the subjects of a first-ever long-duration spaceflight experiment with twins when Scott flew aboard the ISS for a year. Image credit: NASA

while in extended weightlessness. Medical researchers are not yet entirely certain what causes this, and why it appears to be permanent in some astronauts and temporary in others.

More surprising were the changes to his genes. The end caps of many of his genes, structures called telomeres, lengthened while he was in space. Telomeres affect aging in humans, so this was of great interest to the researchers. But most of the telomeres returned to normal within days of Kelly's return to Earth.

About 7 percent of his genes have been impacted more permanently, however, and not in good ways. These were genes related to bone density and formation, the immune system, and DNA repair. Medical studies of these worrying effects on Scott Kelly are ongoing, and the results of this work will be of great concern to planners of space voyages and orbital stations.

Tarver has spent a lot of time working with the Kelly brothers and other astronauts on physiological research, but his work involves psychology as well. On Earth, humans rarely spend months or years at a time locked into small spaces. This is simply not a natural state of existence. For the few that are confined for extended periods of time, there are usually intervals of free movement, as one might experience in an Antarctic research station. Submarines are the major exception.

By World War II, sailors were routinely spending many weeks on patrol underwater with only brief time outside when the vessel was surfaced. A decade later, in the mid-1950s, nuclear-powered submarines could cruise underwater for months at a time. The psychology of being cooped up in small spaces for prolonged periods was becoming ever more important for military health research.[36]

These studies, which continue to this day, reveal that extended confinement to small spaces disrupts human behavior. Sleep cycles are interrupted, working shifts must be adjusted, and caloric intake controlled to ensure peak performance. Other issues, such as territoriality and

Close working quarters aboard a World War II–era submarine. These confined conditions provided some of the US Navy's first data on the effects of working in enclosed spaces for prolonged periods. Image credit: US Navy

Modern submarines, such as this Ohio-class vessel, have far more spacious accommodations for the crew. But with submerged missions lasting for months, psychological issues are still a concern and continue to be studied. Image credit: US Navy

undesirable dominance behaviors, become concerns, as well. As a result of these studies, submarine design and accommodations have been greatly improved.

Unfortunately, spacecraft design is less flexible than that of submarines. Mass is always at a premium, so while efforts are made to accommodate the needs of crew members, interior spaces will always be cramped and luxuries extremely limited. Skylab had a large window installed that provided a psychological boost to the crew; this turned out to be the most popular part of the station for relaxing. The ISS has a larger windowed "cupola" that serves the same function. Where possible, designers select colors to be soothing and attractive. But spacecraft design is highly driven by function, and any mass used for noncritical purposes must be selected with excruciating care. Given these constraints, NASA has utilized available studies—such as the Navy's—and conducted hundreds more of their own to determine priorities.

Part of the medical staff at NASA is composed of psychologists and psychiatrists who monitor and seek to minimize the effects of being confined in a spartan, noisy environment such as the ISS. Any given module of the space station is not much larger than a midsized RV. While there is no floor or ceiling per se, making the total volume usable in any orientation, it can still feel cramped after a few weeks or months.

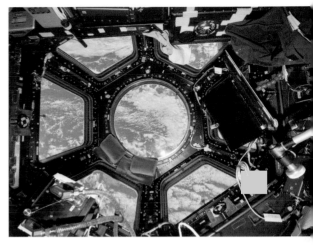

Window structure aboard the International Space Station. This area, also known as the Cupola, is where astronauts spend a large percentage of their off-duty time, watching the Earth move slowly below. Image credit: NASA

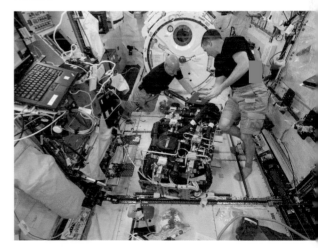

While the International Space Station is the largest structure ever to be assembled in orbit, areas can still seem cramped over time. Image credit: NASA

The Orion capsule, seen here, is second only to the space shuttle in terms of interior space. It is still too small for extended spaceflight, however. Image credit: NASA

NASA-supported simulations, such as this Mars analog station called HI-SEAS in Hawaii, attempt to study the effects of confined quarters on crews. Image credit: NASA

"NASA is working hard on keeping the human psyche healthy in long-duration flight," Tarver says. "There are lots of questions—how much room does a crew of four people need when they are on a mission, for example? Something the size of a compact car is too small for a voyage to Mars, and that's what the Orion capsule is like inside. So how much more volume do you have to attach to it? How much time does an individual person need away from everybody else? How long can you live in those small, artificial environments before you suffer from severe depression?"

NASA's studies have ranged from sealing astronauts in small spacecraft simulators on the ground for weeks at a time to full-on simulations of space stations and Mars bases. No simulation is

The HI-SEAS simulator, with space-suited crew member nearby, is sited near Mauna Loa on the island of Hawaii. Crew members must undergo simulated depressurization and repressurization cycles each time they leave or reenter the station. Image credit: NASA

perfect—they all operate at full Earth gravity, and the participants know they can escape the confines of the facility in an emergency. Nonetheless, a good deal of useful data has been collected relating to enforced, confined cohabitation of cramped spaces.

One of the longest simulations was a sixteen-month Mars-500 study that took place in Russia from 2007 to 2011. A crew of five men—three from Russia, one from France, and one from China—were carefully screened and psychologically profiled before the simulations began. Three "missions" of increasing duration were conducted, with the final, five-hundred-day study closely modeling a round-trip to Mars and back. In this case, the crew was confined to a modular habitat situated inside a building at the Russian Academy of Sciences' Institute of Biomedical Problems in Moscow. There were five interconnected modules and a small simulated bit of Martian exterior terrain—that was the extent of the participants' lives for a year and a half. Communications with the outside world were delayed to replicate the time it can take a message to travel from Mars to Earth and back, up to twenty minutes. Mars-like requirements for exiting and entering the simulator were observed. Crew members had to suit up before entering a model airlock, simulate a depressurization cycle before going outside, and reverse the procedure before reentering the habitat.

Psychological aspects of the extended simulation were recorded and studied, and

The Mars-500 simulator in Moscow, Russia. *Image credit: ESA*

extensive experiments were conducted. Some were physical and some psychological. Results pointed to disrupted sleep cycles, attempts by crew members to isolate themselves from the others, disputes over authority, and avoidance of exercise and group activities.[37]

Flight surgeons and psychologists also have data from a half century of psychological profiling of astronauts before, during, and after flight to draw from. Self-reporting of depression during spaceflight is rare, but depression can show up via long-term monitoring of astronauts during flights. It's a major concern for longer missions. Tarver reports that there have been at least two confirmed cases of depression during orbital missions, and there are likely more that have gone unreported.

Between careful crew selection, improved amenities in spacecraft, and the judicious use of antidepressants when necessary, the mediation of psychological issues during spaceflight is becoming better understood and managed. That's important, because as we go farther into space, dealing with such situations will be up to the crew—on a voyage to Mars, mission control will be far away and communications will be delayed. A careful balance between ground-based intervention and interventions by other crew members, rather than ground-based psychologists, will have to be reached. It has been suggested that at least one member of the crew should have psychological training to deal with such situations on long-term spaceflight in the future.

Then there is the problem of extended weightlessness. If only it were as simple as it is in *Star Trek*—simply push a button and the craft's gravity is turned on. It's never explained; it's just there. But gravity doesn't work like electricity. In space, gravity needs to be generated via the application of centrifugal force. This involves spinning a structure at the end of an arm or tether, which generates an outward force on the swinging object. If this includes a habitat of some kind, the force exerted by this motion mimics gravity effectively for those within. Imagine a Coke can filled with marbles being swung on a three-foot string: The marbles shift to the bottom of the can—the surface farthest from your swinging hand. Make that Coke can a spacecraft, the

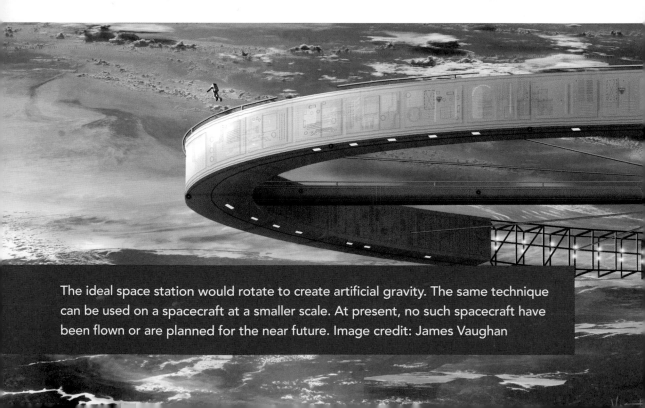

The ideal space station would rotate to create artificial gravity. The same technique can be used on a spacecraft at a smaller scale. At present, no such spacecraft have been flown or are planned for the near future. Image credit: James Vaughan

marbles passengers, and the bottom of the can the floor of the habitat, and you get a general idea of how this would work. This is exactly how the centrifuge in the movie *2001: A Space Odyssey* was depicted.

There are limits—what speed the spacecraft or habitat can be spun, how far from the attachment point the floor of the habitat rests, and other engineering considerations. For the sake of this discussion, let's assume that we are looking at a modestly sized spacecraft that is, like all spacecraft to date, designed to accomplish as much as it can with as little mass as possible. So this will not be a huge, spinning centrifuge as we have seen in such movies as 2016's *Passengers*; for the near future, a space centrifuge is likely to be of more modest proportions. It's also likely to be generating just a percentage of Earth-normal gravity.

Just how much gravity is needed to keep humans healthy has not been determined. The only time astronauts have experienced partial gravity was during the moonwalks of the 1960s and 1970s. The moon has about 17 percent of Earth's gravity, but the astronauts' stays there were so brief that its effects on health were impossible to measure. A low-gravity centrifuge in space could allow us to observe the impact, both negative and positive, of lunar gravity and Martian gravity over an extended period.

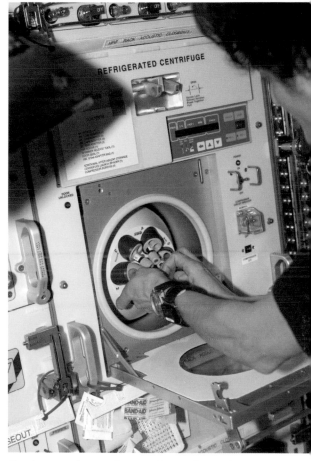

To date, only small centrifuges have been flown on the International Space Station. Image credit: NASA

At the moment, we only have a tiny, experimental centrifuge in space that can be used to study the effects of partial gravity. The first experiments will be with fruit flies—they breed quickly and will give some fast baseline results. Rodents will be next, in a larger centrifuge, to gauge partial gravity's effects on mammal tissue.

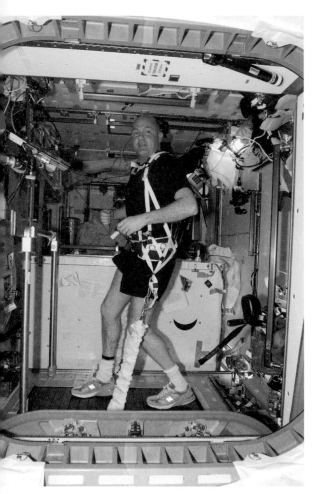

Astronaut André Kuipers exercises on the COLBERT space treadmill aboard the International Space Station. At least two hours per day are required to keep astronauts healthy. Image credit: NASA

available and make do with other countermeasures to offset the effects of microgravity. At this point in time, this is accomplished by vigorous exercise, which has proved to be fairly effective. But it has one big drawback—time.

Astronauts on the ISS require at least two hours of rigorous exercise per day just to keep their bodies fit. "That's inserted into the eight hours of work that they need to schedule," Tarver says. "There are two different exercise sessions every day. One is a treadmill or a bicycle, and the other one is the advanced resistance exercise device, like lifting weights in one-g. That's just to replace the effects of walking around on the planet."

At least two hours of exercise per day just to retain basic fitness is a lot of lost productivity—about a quarter of an astronaut's working day—and time is an expensive commodity in space. Tarver and others believe that, in the future, centrifuges will prove critical for maintaining health and maximizing efficiency during extended spaceflight.

Finally, we come to the concerns about extended exposure to radiation. The only experience we have with humans traversing the hostile regions outside Earth's protective magnetic field dates back to the Apollo missions, and these flights exposed the crew to deep-space radiation for less than twelve days. The ISS is in low Earth orbit and is therefore relatively protected

Then centrifuge units, not yet built but on the drawing boards, would test the effects of partial gravity on humans.

Until such machinery can be flown, we must use the limited test results now

from radiation's searing effects, so while levels there are higher than on Earth's surface, they don't come close to those found beyond the Van Allen belts, the magnetic field that surrounds Earth, which extends out to about 36,000 miles. On a long journey to the moon or Mars, or an outpost on either world, humans will be exposed to high levels of radiation over time. The same goes for long missions to stations in cislunar space or to asteroids.

There are two sources of radiation in space: the sun and stars beyond our solar system. Solar radiation includes high-energy protons that are dangerous to humans. The sun's radiation output varies, running in roughly eleven-year cycles that are fairly predictable. Occasional "sun-storms" can magnify this dangerous output significantly. These will have to be monitored constantly from Earth, with warnings sent to any crews operating outside of Earth orbit to seek appropriate shelter when they occur.

The radiation that comes from outside the solar system is even more damaging. It consists of a constant flow of high-energy protons and electromagnetic energy. These are called galactic cosmic rays (GCRs).

The Van Allen belts are Earth's first line of defense against space radiation. They extend out to about 36,000 miles. Image credit: NASA

A cluster of massive stars, surrounded by clouds of interstellar gas and dust. This cluster is located 20,000 light-years away in the constellation Carina and contains the type of huge, hot stars that identified as a typical source of cosmic rays. Image credit: NASA

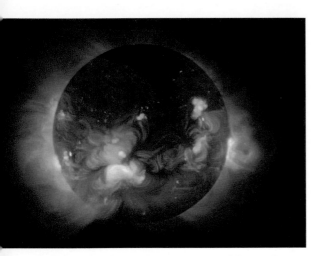

A map of solar radiation created from data returned by a NASA space telescope. Image credit: NASA

This galactic radiation penetrates the Earth's magnetic field more readily than solar radiation does, so our bodies are conditioned to receive a certain dose. But combine both these types of radiation in deep space, and the result is a toxic cocktail of radiation exposure that can create cellular damage within months. The research is far from conclusive, but possible dangers include cancer from scrambled DNA, as well as cardiovascular issues, musculoskeletal problems, and neurological damage.

NASA began studying space radiation before it first sent astronauts into space in the 1960s. Since then, spaceflight guidelines regarding lifetime exposure limits to space radiation's potentially deadly effects have been established. "We've had astronauts flying in space for fifty years,

and on the ISS for about twenty," Tarver says. "Some of them have flown multiple missions, and we realized that they could eventually bump up against NASA's administrated 'red line' for lifelong radiation exposure."

For context, think of it this way: When you get a medical X-ray, you are put into a machine that bathes you in a brief dose of high-energy radiation. So long as you only do this once a year or so, your lifelong dose is not considered risky. But have you ever noticed how the tech who runs the machine leaves the room before activating it? They are minimizing their overall exposure. They work around that machine day in and day out, and if they stood there as it imaged your bones or teeth, they would eventually run a much higher risk of cancer and cellular damage. It's the same with astronauts—short missions, especially those in low Earth orbit, are minimally dangerous with regard to radiation, but long flights increase this risk. Notably, once on the surface of Mars, the amount of galactic radiation is cut by about half, but solar radiation is still an issue. The moon, with no atmosphere, offers no protection from either type.

Just how powerful is radiation beyond Earth orbit? The Mars Science Laboratory rover Curiosity, launched in 2012, carries a radiation-measuring tool called the Radiation Assessment Detector (RAD) as part of its instrument package. While

The Radiation Assessment Detector (RAD) instrument aboard the Curiosity Mars rover measured radiation between the earth and Mars, and on the Martian surface. Image credit: NASA

this instrument was intended primarily to determine the amount of radiation striking the surface of Mars, it was turned on during the nine-month transit from Earth to the Red Planet, and it delivered a lot of useful data even before landing. As summarized by Cary Zeitlin, one of the scientists responsible for the instrument, the radiation environment in space "is several hundred times more intense than it is on Earth and that's inside a shielded spacecraft."[38] This is not good news for future space travelers. A trip to Mars and back would dose each astronaut with about two-thirds of a sievert, about fifteen times as much radiation as that allowable for workers at nuclear plants. One sievert raises the risk of cancer development by about 5 percent. While not necessarily fatal—especially when compared to the other risks faced by Mars voyagers—these levels of radiation exposure are noteworthy and an ongoing area of concern.[39]

Bill Tarver explained the rationale for the radiation exposure rules in more detail: "This dosing limit is based on what radiation workers experienced in atomic submarines and nuclear reactors back in the day, which resulted in about a 3 percent increase in cancer-related deaths across their careers. NASA said 'we're not going to do that.'" But the rules are derived primarily from Earth-based studies; we simply do not have enough experience operating with humans beyond low Earth orbit to gather much data that way. Mars missions, lunar bases, or even space stations out near the moon would meet or exceed the current limits quickly if the inhabitants were not protected.

When SpaceX starts flying its Big Falcon Rocket in the 2020s, the passengers—as many as one hundred of them—will need to be shielded from space radiation during longer voyages in space. Image source: SpaceX

Most discussions about space radiation and astronauts focus on genetic damage resulting in increased cancer risks across their lifetime. But there may be other dangers from radiation exposure. Research at the University of Rochester Medical Center from 2012 looked at the effect of exposure to high-energy particles on the brains of mice using the Brookhaven National Lab's particle accelerators as a radiation source. The dosing was estimated to be about equivalent to what a human might experience on a three-year Mars journey. After the mice were irradiated, researchers put them through a series of memory tests. The mice that had been bombarded were significantly worse at tasks that involved memory and location recall. They also displayed symptoms of Alzheimer's disease.[40] More research needs to be done to validate these results.

All spacefaring powers are concerned for the people they send into space. Russia and more recently China have established their own sets of radiation guidelines. But these standards will soon be pressed—not just by the requirements for longer missions, but also by the changing nature of spaceflight. As the private sector begins to send people beyond Earth orbit, who will dictate the safety protocols? How much risk of debilitation from extended weightlessness or radiation exposure should a private citizen in space, flying with a company like SpaceX, for example, be allowed to assume? These questions will need to be addressed soon, as private entities begin sending paying passengers to places like the moon.

Solutions to the problem of radiation exposure are being sought through experiments conducted by NASA and other space agencies, as well as universities worldwide. Many approaches that appear to be both effective and workable are being researched, and the results are promising, but all involve some amount of added spacecraft mass that would have to be lofted to orbit, the bane of affordable spaceflight. The trick to getting it right is to find the optimal combination of effective protection and low mass.

In Earth orbit, just a few millimeters of metallic hull—aluminum is the most common metal used in spacecraft and the ISS—is sufficient to shield humans from much of the radiation that makes it past Earth's magnetic field. But once the spacecraft ventures deeper into space, the requirements expand dramatically. To make things more complex, the radiation from the sun requires a different kind of shielding than the radiation from deep space.

The concrete casing used to shield nuclear reactors is much too heavy to be used in space, so we need something lighter. It's tempting to think that putting humans inside a thicker metal cylinder would alleviate radiation exposure, but that's not entirely true. Aluminum and steel are not great at stopping the worst of the radiation hazards, and in some cases can magnify them. The impact from an energetic particle from space can knock more particles off a metal hull's interior, and these can amplify the radiation's negative effects.

Various other schemes have been proposed to mediate radiation risks in space. Currently, the most popular ideas for deep-space shielding involve wrapping the crew compartment in something light and nonmetallic that will absorb and scatter radiation, such as plastic sheeting or tanks filled with liquids, since both water and fuel are useful for mitigating radiation. These shields will not magnify the effects of a particle impact the way metal does. Sandwiched layers of plastic and metal are another proposed solution, but

SKYLAB WORKSHOP
CREW QUARTERS

Traditional aluminum hulls, such as those used on the International Space Station and Skylab (seen here) are fine for low Earth orbit, but for travel between the planets and stations orbiting other worlds, better shielding will be needed. Image credit: NASA

the materials would have to be fairly thick to be effective.

The amounts of water or fuel required for shielding are also large—estimates range from four to ten feet of either, in order to protect the crew. But rocket engines use a lot of fuel, so storage tanks could surround the crew compartment and provide a reasonable amount of protection. A 2005 NASA study suggested that tanks of liquid hydrogen could be quite effective, for example, if coupled with radiation-inhibiting fuel tanks—palladium alloys were mentioned—and would kill two birds with one stone, so long as there was fuel in the tanks.[41] Large amounts of water would accomplish something similar.

The liquid shielding method equates to a lot of launch mass—not just the structure itself, but all the liquid needed to fill it. Water is particularly heavy. Unlike plastic sheets, though, water and liquid hydrogen are useful in space. Spacecraft must carry fuel and water anyway, so up to a point, this large volume of liquid shielding also provides a source of fuel, drinking water, and breathable air. You just need to keep enough of it handy to provide the necessary protection. But it's still very expensive to launch all this mass into space.

Just how expensive? By NASA's own estimates, it used to cost about $80,000 for the shuttle to launch a single gallon of water into orbit; SpaceX can now do

Deep-space tankers and depots, storing resources mined from asteroids and the moon, will supply water and hydrogen that can be used for fuel and shielding. Image credit: James Vaughan

the same for far less, about $11,000.[42] In either case, you can imagine that filling a huge orbiting tank with water or even liquid hydrogen (which is lighter and therefore cheaper to launch per gallon) would be costly. There must be a better way.

As it turns out, there is. The foregoing assumes that we would need to launch all the materials needed for a Mars mission from Earth—the water or liquid hydrogen for shielding alone could require many launches. This does not even include the rest of the spacecraft or the astronauts, which will require another two or three launches. But what if you could find the water and hydrogen needed to accomplish this in space? This could change the discussion significantly. Ongoing work toward mining water from the moon and asteroids would pay off handsomely for shielding mass—there is a lot of water on the moon and in asteroids, just waiting for us. This will take years to accomplish but would be a game changer.

There have also been more imaginative suggestions. Some designers have proposed using a hollowed-out asteroid as a spacecraft—capture it, drill a hole into the middle for your crew quarters, attach some engines to it, and travel the cosmos in well-shielded, rocky comfort.[43] While this is a possibility in the far future, the technology required to provide sufficient propulsion to move an asteroid ship from one place to another is not currently available. The idea has promise, however; the same asteroid used as a spacecraft could contain an abundance of water ice, which can be mined for fuel and life support. Experiments are also underway to explore other alternatives, such as lining craft hulls with a thick layer of lithium hydride powder that effectively blocks most kinds of radiation.[44]

These radiation-shielding scenarios are specific to transporting humans from one world to another. Once a crew has arrived at a distant body in space—let's say the moon or Mars—other potential solutions are available. Habitats can be buried under a few yards of soil, or built in caves or lava tubes. Any of these scenarios offer protection and are ideal since

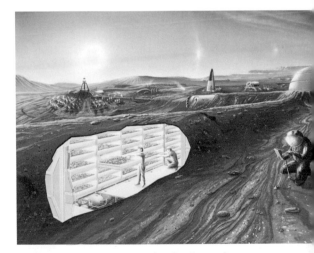

Colonies on Mars may be built under a layer of Martian dirt, also called regolith, or underground in tunnels, such as lava tubes. Image credit: NASA/Ames

much of the necessary building material is already there. All you need to do is move or excavate it.

Scientists are studying at least three types of active radiation shielding. What the different types all have in common is that they involve the generation of some sort of electromagnetic field radiating outward from a spacecraft's hull. This field repels or deflects incoming radiation.

The first type of active shield would generate an electrical field with a negative charge around all or part of the spacecraft.[45] This would repel and redirect most negatively charged particles—solving only part of the problem—and would have to be used in conjunction with passive shielding technology. There are no known downsides to exposing the humans inside the spacecraft to this kind of electrical field for the durations required.

A second approach involves generating a strong magnetic field around the spacecraft, essentially mimicking the protection provided by the Earth. This would require massive superconductors and the related technology necessary to keep them cold enough to operate properly. Because exposure to strong magnetic fields over long periods has some negative effects on humans, the protective shields would have to be formed outside of and around crewed quarters, but not allowed to impinge on the humans inside. This is challenging but appears to be doable.

Magnetic shielding is thought to be particularly effective against galactic cosmic radiation (GCRs).

The final approach can best be described as plasma shielding and utilizes a mass of ionized particles trapped within electromagnetic fields. Plasma fields have both strong magnetic and electrical properties and might be the best way to deflect and redirect radiation. Plasma shields are thought to require the least mass and energy to accomplish, but they also involve the least-developed technology.

Such "deflector shields" are very appealing in theory. There are drawbacks to creating them with currently available technology, however. Such field generators are very power hungry and would require either vastly improved solar panels or, more likely, a nuclear power source. NASA is studying a new spaceflight-capable nuclear reactor design, building on work undertaken in the 1960s and 1970s. Improved nuclear power sources would greatly enhance the possibility of using active shielding on deep-space human missions.

A combination of active and passive shielding might provide the best approach. The active shield would be projected around all crewed areas of the spacecraft, with heavier, passive shielding—which cannot malfunction, since it is essentially inert or a contained liquid—used for smaller "storm shelters." With this design, the active shield need only be effective for general doses of radiation, and the passive shelter could be used during malfunctions of the active shielding or during a solar flare, when radiation levels would spike temporarily.

Humans will eventually travel across interplanetary space; it is an inevitability. How the dangers of this journey are mitigated is being extensively investigated today. With enough research, and the application of the right blend of technology, caution, and biomedical intervention, we should be capable of protecting people for the duration of a Mars trip within a decade.

NASA, numerous universities, and international space agencies are working hard on the problems encountered during long-term space travel and will find viable and robust solutions given time. The next step is getting that technology—and the spacecraft that will utilize it—into space inexpensively and safely. This is increasingly becoming the realm of a new entity: the spaceflight entrepreneur.

Some designs call for water to provide radiation shielding not just for spacecraft, but habitats as well. This NASA concept for a water-ice shielded Mars habitat would accomplish this with a thick layer of frozen water derived from Martian resources. Image credit: NASA

MARS: OMAHA CRATER
GMT: 14:00

Image credit: James Vaughan

CHAPTER 7

SPACE ENTREPRENEURS

The two-hour drive from Los Angeles to the city of Mojave, population 4,238, is a dreary one. I had set out at 7 AM on a cloudy Friday morning to cross this forlorn desert, an expanse I've seen perhaps forty times but still have not quite gotten used to. Deserts are supposed to be places of desperate and forbidding beauty—huge, sweeping fields of sand dunes, stately saguaro cacti, and the occasional scrawny coyote prowling the hardpan or lone eagle silhouetted against the sky. Not so, this region of Southern California. It is, for the most part, a flat, cocoa-brown wasteland with little vertical relief and only scruffy, wizened shrubs, with the arid Tehachapi Mountains in the distance. The transition from the Los Angeles metropolis to high desert is gradual. Once past the extended suburbs of Palmdale and Lancaster, one passes through sun-bleached housing developments and generic industrial parks until the city runs out, as if it simply lost the will to go on. Only the occasional cluster of gas stations and fast-food outlets break the desolation.

93

The drive to the city of Mojave, California. Image credit: Wikimedia Commons/Theschmallfella

Then you cross into the historic city of Mojave, which began life as a railroad town in 1876 and later became a stopover on the main highway between Southern California and the old gold-rush country to the north. In 1964, the construction of Highway 58 largely bypassed the town. The business district is now a combination of gas stations and fast-food franchises interspersed with older, sometimes abandoned buildings, from which paint peels like cornflakes.

I drove about halfway through the nine-block length of the town to Airport Road and turned right. Here the scenery improves rapidly. A new energy buzzes in old Mojave. Soon the small airport became visible—a cluster of metal buildings surrounding a modern control tower. Mojave Airport serves as a hub for business activity for the region, and within its confines stands Virgin Galactic, where billionaire Richard Branson is slowly realizing his spaceflight dream.

Branson has been working on a plan to democratize space tourism since 2004; he has named his newest spacecraft

Unity. This modern craft is patterned after a decade of pioneering work by Burt Rutan, the man who built a rocketplane to win the Ansari X Prize. Rutan won the multimillion-dollar cash award in 2004 for flying SpaceShipOne twice in one week to the edge of space. He subsequently merged his efforts with Branson, though Rutan later left the company to pursue other space ventures. But his unique and innovative rocketplane design concepts live on.

Entrance to Mojave Spaceport. Image credit: Wikimedia Commons/Californiacondor

Virgin's assembly facility fills a vast modern building amid smaller, older hangars. The only way you would know that anything special takes place here is the large sign across the single glass protrusion out front bearing the word *FAITH* in tall white letters—an acronym for Final Assembly, Integration, and Test Hangar. From the outside, the facility takes more cues from the rough-and-ready days of test flight than from NASA's massive operations in Florida and Texas. Just a few miles down the road at Edwards Air Force Base, the X-15 tore up the skies in the 1960s. Virgin Galactic seems infused with that pioneering spirit.

I park in a half-empty lot and enter for a tour of the plant. Inside, Virgin Galactic is all business, with little effort wasted on glitz. The double doors swing open and I am greeted with the smells of a classic machine shop—oil and solvents. The workers are young, idealistic, and friendly. After a few brief introductions, they return to building their rocketplane. A few machinists craft parts here and there, and many more folks look intently at the computer monitors haphazardly strewn on banquet tables spread across the workshop floor. Engineers and technicians gather in small clusters, speaking in low tones. If it weren't for the magnificent rocketplane and its carrier craft sitting boldly in the middle of the shop, gleaming white and chrome under the bright lights, it could be any midsized machine

The author inside Virgin Galactic's fabrication and assembly facility, in front of the carrier airplane *Eve*. This twin-fuselage jet carries Virgin's rocketplane *Unity* to altitude before it detaches and begins its rocket-powered climb to space. Image credit: Rod Pyle

Virgin Galactic's Mojave facility. Image credit: Wikimedia Commons/Ed Parsons shop. To someone used to the vast spaces and precise, Germanic layouts of NASA facilities, this looks almost casual—just a couple dozen workers on any given shift, enthusiastically fabricating the future.

A group of engineers—three men and two women—are collected underneath *Unity*'s left wing. The rocketplane is supported on jack stands, and the left landing gear is up, tucked inside the fuselage as it would be during flight. They are about to conduct a gear-down test, simulating the process by which *Unity* will lower its wheels prior to landing. After a bit of fussing, and with nods of agreement, the group moves back to a safe distance. Someone says "clear!" and with an alarmingly loud *bang* the landing gear is forced into a down-and-locked position via high-pressure gas. There are smiles all around, and the crew moves on to their monitors to evaluate the test data. It's a moment of muted excitement and welcome progress. Virgin Galactic is behind schedule—the initial passenger flight was supposed to take place in 2009—so progress is critical.

Virgin's approach to launch is unique to commercial passenger-carrying rocketplanes. It will be flown to launch altitude of about 52,000 feet in between the twin hulls of a giant carrier plane, WhiteKnightTwo, or *Eve*, which uses traditional jet engines. *Unity* will then ignite its rocket engine to fly into suborbital

An X-15, with extra fuel tanks attached, is carried to altitude by a USAF B-52 bomber before dropping free and igniting its rocket engine. Virgin Galactic's rocketplane operates in a similar fashion. Image credit: US Air Force

space. This launch procedure is based on military rocketplanes from the past such as the X-15. Flying in the 1950s and 1960s, the X-15 was carried to altitude by a B-52 jet bomber before igniting its rocket engine and soaring to the edge of space. Using a carrier plane saves fuel for the rocket engine and provides a huge speed and altitude assist at launch.

Virgin Galactic's plans to carry paying passengers have been consistently delayed. In a ground test in 2007, a rocket engine exploded, killing three engineers. The engine design has since been improved,

Virgin Galactic's newest rocketplane, Unity, *mated to its carrier plane* Eve, *at its first rollout. While it does not return at orbital speeds,* Unity *still needs extensive thermal protection to survive the high temperatures of reentering the atmosphere. Image credit: Virgin Galactic/Mark Greenberg*

years came and went with only intermittent testing occurring. Clearly, creating a space-tourist business was harder than Branson, or any of the others working in the field, had foreseen.

Virgin Galactic hired a new chief executive named George T. Whitesides in 2010. Whitesides is a veteran of other spaceflight activities, including a brief stint as chief of staff at NASA. He was the executive director of the National Space Society from 2004 to 2008. Whitesides and his wife, Loretta, also started a yearly space event called Yuri's Night, a celebration of the dawn of the space age, that has spread globally, resulting in thirty-five events in fifty-seven countries. His father is a prominent chemistry professor at Princeton University who has done

but despite numerous test flights, it has still suffered teething problems. One of the biggest complications has been getting the thermal protection on *Unity* right. Despite being a suborbital rocketplane, it still encounters high temperatures during its descent back into the atmosphere.

Virgin Galactic's history is indicative of the perils facing spaceflight entrepreneurs. In 2008, despite the accident, Branson projected that tourist flights would begin by the end of 2009. Then, when that date approached, the company announced that it might take two more years to begin operations. Those

Virgin Galactic's founder, Richard Branson, greets the press at the rollout of Unity. *Image credit: Virgin Galactic*

George Whitesides, president of Virgin Galactic. Image credit: Virgin Galactic

discouraged his boss from setting unattainable operational goals.

By 2013, the workforce had expanded and the company had flown dozens of test flights: fifteen tests with the rocketplane attached to the carrier craft, and sixteen more tests in which the rocketplane separated and glided to a landing. Within the year, two powered tests were flown, with *Unity*'s predecessor performing properly, but not quite attaining the desired targets for speed and duration of flight.[46]

Then, in 2014, with almost fifty tests behind them, an in-flight accident occurred during a powered flight of SpaceShipTwo with two pilots aboard. On October 31, 2014, SpaceShipTwo was dropped from the carrier plane above the Mojave Desert. The engine fired and the rocketplane began to climb. Eleven seconds later the ship malfunctioned and broke up. One pilot survived with injuries, but the other perished in the accident. Despite conjecture about a possible explosion of the sometimes-troublesome rocket engine, the cause of the crash was later determined, after a long and arduous investigation, to be the reentry mechanism. Virgin's rocketplanes have a unique "feathering" system that moves the tail surfaces relative to the main hull during the early stages of descent. This system was apparently—and accidentally—deployed early in the test flight, while the rockets were still firing, rather than after

extensive research in materials science and nanotechnology, so science and tech run in the family.

Once Whitesides settled in at Virgin Galactic, he "adjusted how they did business," as he put it in an interview for the *Guardian* in 2014. "We've changed dramatically as a company . . . When I joined in 2010 we were primarily a marketing organization." Whitesides refocused the emphasis on engineering and testing, and

shutdown, as designed. The craft disintegrated due to violent aerodynamic stresses.

In a press release highlighting its final write-up of the accident, the National Transportation Safety Board (NTSB) noted that there appeared to have been human error involved in the early deployment of the feathering system.

"The National Transportation Safety Board determined the cause of the Oct. 31, 2014 in-flight breakup of Space-ShipTwo was Scaled Composite's [Rutan's original company] failure to consider and protect against human error and the co-pilot's premature unlocking of the spaceship's feather system as a result of time pressure and vibration and loads that he had not recently experienced."[47]

The report continued to detail that the pilots had not been properly trained, and that the feathering system should have had a safety-interlock system, which has since been added. Some blame was apportioned to the FAA as well—it was suggested that the agency had been too generous in granting operational waivers and exceptions for experimental flights. The report then provided a summarizing statement that might best describe why entrepreneurial efforts, such as Virgin Galactic's, take longer than anticipated: "Manned commercial spaceflight is a new frontier, with many unknown risks and hazards. In such an environment, safety margins around known hazards must be rigorously established and, where possible, expanded." The report concluded, "For commercial spaceflight to successfully mature, we must meticulously seek out and mitigate known hazards, as a prerequisite to identifying and mitigating new hazards."

This second accident, after the 2007 engine explosion, shook Virgin Galactic to its core and resulted in further strong and proactive steps to ensure future safety. More test flights were scheduled to ascertain that each system and subsystem was safe and reliable.

The entire private spaceflight industry also took notice; accidents can happen, but when human life is at risk, you must act promptly, and even more urgently when those lives are civilian passengers. Just because you can get something approved by a government entity like the FAA does not mean you can stop obsessing about safety. You must go above and beyond to ensure that your system is as trustworthy as you can make it.

Many existing regulations have been built around decades of activities by NASA and the military, and they have yet to fully catch up with the private sector's new spaceflight efforts. And while new regulations are needed, spaceflight will never be 100 percent safe. As American author John A. Shedd famously said in 1928, "A ship in harbor is safe, but that is not what ships are made for." To move outward, we must leave the harbor, an inherently risky business. As space entrepreneurial efforts proceed, especially those carrying humans, more risk will be assumed, and we must collectively learn to accept it, within the regulations, rules, and processes that will be established to mitigate and minimize that risk.

For any commercial spaceflight company to succeed, it must be able to provide frequent, reliable, and affordable access to space while generating a profit. But the challenges of conducting safe spaceflight in an airline-like business model are substantial. You can see why this is a realm best suited to billionaire investors—it takes a lot of time and money to accomplish, and no revenue is generated during the development and testing period. Profits are still in the future for Virgin. The word *FAITH* on the hangar's exterior would appear to be a key attribute of people like Branson. He and his partners have been investing for almost two decades and have shown incredible tenacity in their continuing pursuit and development of tourist spaceflight.

Virgin Galactic has moved on since the accident in 2014, and *Unity* is now engaged in an aggressive test-flight program. The company has also expanded its efforts to include launching unmanned cargoes—satellites and other small payloads—with new systems that build on their unique technologies.

The first 2018 test of Unity, *Virgin's new rocketplane. Image credit: Virgin Galactic*

This satellite-launching spin-off company, called Virgin Orbit, was initiated in 2015 with the opening of a separate facility in Long Beach, California. The goal is to fly midsized satellites on a rocket called LauncherOne, which will be carried to launch altitude by a converted Boeing 747, much as *Unity* is carried to altitude by *Eve*.

I spoke to George Whitesides at length about Virgin Galactic's plans and the trials before them. "Space is a challenging technical endeavor and it is very unforgiving given the current technology," he said. "We're trying to do hard things in a world that has more risk aversion than fifty years ago."

When asked about the overarching plan for the company, Whitesides said, "I'm really hopeful that over the course

Artist's concept of Virgin Orbit's carrier plane, a modified 747, lifting the uncrewed satellite launcher to altitude. Image credit: Virgin Orbit

of the next decade we can send hundreds, probably thousands, of people into space." He noted that for the last sixty years, since the beginnings of the first space age, on average, about ten people have been sent into space during a year. "That's gotten lower since the end of the shuttle program," he added. "So that would be a huge shift—we can go to a world where the average person knows someone who has gone into space."

Until then, Virgin will continue test flights to be certain that it has the safest possible rocketplane in which to carry paying passengers. On April 5, 2018, the company completed its first self-powered test of *Unity* after two years of ground and unpowered drop tests. *Unity* was carried to altitude by *Eve*, then dropped free of the carrier plane at 46,500 feet. *Unity*'s

rocket motor ignited, and it climbed at an 80-degree angle, flying at almost twice the speed of sound. It reached an altitude of about 85,000 feet before descending to a successful landing on Virgin's Mojave runway. It was an important milestone for the company, bringing it a step closer to passenger flights.

Virgin Orbit is moving ahead quickly, as well. With far smaller, nonhuman payloads, there is much less at stake. It's an easier path to tread than space tourism. If a satellite is lost, it will have an impact on the companies that create and insure it, but nobody will perish.

But launching satellites is also a business with more current and potential competition than space tourism. Today, Virgin Galactic's only real rival to carry civilians into space is Jeff Bezos's Blue Origin, which is also building and

Unity's rocket engine firing during its first powered test flight. Image credit: Virgin Galactic

testing spacecraft that will shuttle tourists. With satellites, though, there are many up-and-coming launch businesses. Virgin Orbit will need to move aggressively to stake its claim in this marketplace.

Virgin Orbit's biggest current rival is Orbital ATK, a company that has been operating in the air-launched "smallsat" business for almost thirty years and was acquired in 2018 by Northrop Grumman. Rebranded Northrop Grumman Innovation Systems, its small satellite launcher, called Pegasus, is dropped from a carrier plane just as Virgin Orbit's will be. The Pegasus can haul just under 1,000 pounds to low Earth orbit. But it is expensive, as much as $50 million per launch, and Whitesides thinks that Virgin will be able to compete aggressively and lower that price dramatically.

Innovation Systems will not be Virgin's only competition for long, however. Virgin Orbit is just one of many new companies entering the smallsat launch arena. Nipping at its heels are dozens of other contenders intending to launch small payloads at reduced prices. It's a much simpler point-of-entry into the private spaceflight business than flying people. It's also much easier from a regulatory and insurance point of view, with far lower barriers to entry for new entrepreneurs. There's a big market for these commercial ventures as well. A 2016 analysis reports that about 3,000 small satellites may be queueing up for launch between now and 2022, and this does not include the large constellations of broadband-supplying satellites that will be launched by SpaceX and others. Estimated counts of these small broadband providers range from as few as 2,000 to over 10,000 satellites.[48]

There are too many small companies to detail all of them here, and many of them in existence today will not establish themselves as going concerns. But there are a number of serious contenders already conducting test launches of well-financed hardware that deserve mention.

Stratolaunch is the heavyweight in this category. Its planned rocket will be able to launch larger satellites, as hefty as 10,000 pounds, or clusters of hundreds of smallsats. The company is owned by Microsoft billionaire Paul Allen, who hired Burt Rutan after he left Virgin Galactic, and also operates out of Mojave. Stratolaunch will use a huge carrier plane, like the Virgin LauncherOne concept, but built with advanced lightweight composites, to lift a three-stage rocket to high altitudes. The carrier plane is a behemoth, with a 385-foot wingspan and six jet engines salvaged from 747 aircraft. Commercial operations are expected to begin in 2019 or 2020.

Other companies are aiming at smaller single payloads, about 1,300 pounds or less. With rapid advances in

An Orbital ATK Pegasus satellite launcher drops free of its carrier plane. In moments, its rocket engine will fire, propelling it into orbit to deliver a satellite. Image credit: Northrop Grumman Innovation Systems

technology, the lower limit of this weight category is falling quickly—smallsats are getting even smaller. A few years ago, people in the business just called a satellite a satellite. Then, as satellites shrank due to more advanced, miniaturized technology, the term "smallsat" was coined. Soon there were "cubesats," even smaller designs shaped like a box, about four inches per side, with standardized payloads weighing about three pounds. Then the term "nanosat" entered the vernacular, covering anything from two to twenty-two pounds (and confusing the terminology somewhat). There are even "femtosats," which weigh just a few ounces. It's a quickly evolving field.

These small satellites increasingly use off-the-shelf technology, with some experimental units being designed around cell phone computer processors. With their sides covered in little solar panels, even the tiniest cubesats are capable of performing powerful work in space. This includes orbital imaging for commercial purposes, weather tracking, and even scientific research. As sizes continue to diminish, and associated weights drop, dozens—and up to hundreds—of these units will be launched from a single small rocket, increasing the ability for more and more companies to enter the marketplace and for competition to increase.

Among the companies working to establish themselves in the smallsat market, Firefly Aerospace is developing a rocket capable of orbiting 2,000 pounds to low Earth orbit. (All these launch systems will be quoted to low Earth orbit.) Firefly's two founders previously held positions at other space companies—in this case, Virgin Galactic, Blue Origin, and SpaceX. Firefly began operations in Hawthorne, California, in 2014, and has since moved to Texas.

Rocket Lab was founded in 2006 and is located in both Southern California and New Zealand. Its Electron rocket is expected to be capable of launching 330-pound payloads to a polar orbit, at about 300 miles altitude, going over Earth's poles, at a cost of about $5 million.

Polar orbits offer twenty-four-hour coverage as the Earth rotates below the satellite's orbit. The company has flight-tested its rocket, and completed a launch facility in New Zealand in 2016.

Vector Space Systems, a relative newcomer, was founded in 2016 by Jim Cantrell and other aerospace and software industry veterans. Cantrell was associated with SpaceX early on. Located in Tucson, Arizona, the company made quick progress, launching its first test vehicle in 2017. It plans to launch out of traditional sites such as Cape Canaveral, as well as various other locations, using smaller mobile facilities. Vector's rocket is designed to carry 145 pounds to orbit, and to do so more than a hundred times per year.

This is just a sampling of the major providers in the private smallsat launch business. These fledgling companies have in common relatively low costs and the ability to launch quickly once an order is

A Vector test launch in 2017. Image credit: Vector Space Systems

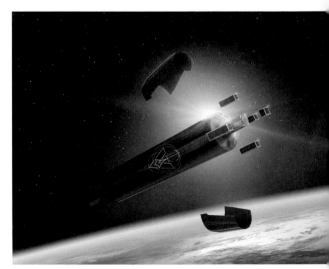

Artist's concept of a Vector launch, with cubesats being deployed in orbit. Image credit: James Vaughan

placed. Still, they face some similar challenges to the big payload companies: funding, launch-facility regulation by the government, the rigors of designing small, reliable rocket motors, and guiding those rockets—and the payloads they carry—into the proper orientation in space.

So far I've talked about launch providers. These rocket builders are the George Clooneys and Taylor Swifts of the space business, with lots of flash and dazzle. But there are countless other enterprises working in less obvious categories in which space entrepreneurs are making a difference—and, in some cases, hefty profits.

Perhaps the most notable of these is a company called NanoRacks, founded in 2009 to fill a niche in which few had seen profit potential—an area called standardized flight interfaces. Until a few years ago, satellite hardware was custom designed, unique to its builder, and had to be created to fit a given rocket. This is an expensive and complex way to do business. NanoRacks was founded to provide a standard interface between satellites and the rockets they fly on. A very simple metaphor is the charging jack on newer smartphones—most now use a USB-C connector. But just a few years back, smartphones used many different types

A NanoRacks assembly deploys two Planet Labs Dove cubesats from the International Space Station. Image credit: NASA

of power connectors, and it was a nightmare for consumers. Like the USB-C connector on a cellphone or computer, the NanoRacks launch interface simplifies and regiments the link between satellites and rockets. NanoRacks has taken this a step further by providing standardized cubesat deployment mechanisms that work with the International Space Station's robotic arm.

NanoRacks' CEO Jeffrey Manber previously worked on various space enterprises, including a stint as a business ambassador between NASA and the Soviet Union's space program. The company's short history is impressive, with about six hundred payloads handled to date. Their client list includes companies small and large, with payloads ranging from national security satellites to tiny cubesats engineered by university students. They also provided some of the first commercial hardware to fly on the ISS.

Planet Labs is another Space 2.0 success story. The company was founded in 2010 by a pair of NASA engineers to design, manufacture, and fly small satellites that would be capable of imaging Earth at a fraction of the cost of existing satellites. The company's designs are based on the cubesat form factor. Planet Labs' satellites are called 3U (3-unit) cubesats, and they are about four by four by twelve inches in size. They don't require dedicated launches as traditional Earth-imaging

Planet Labs cubesats ready to go. Image credit: NASA/Planet Labs

satellites do, and can be included as secondary payloads—sometimes on NanoRacks hardware. Today, a swarm of these cubesats creates a complete image of the Earth's surface each day, available by subscription via a web-based interface. With almost two hundred satellites currently in orbit, their "fleet" is currently the largest ever flown by one company.

Made In Space, Inc. is another interesting example of space entrepreneurism operating in a newly discovered market niche. The company was started in 2010 by four students who met at a Silicon Valley training program. The quartet created the first space-rated 3-D printers in a small lab set up in the NASA Ames Research Park in Mountain View, California. By 2011, they were flying their experimental printers in NASA-provided zero-g simulations, and in 2014, after years of exacting design and testing, they flew a 3-D printer to the ISS—the first such machine to enter space.

The technical constraints of working on a highly regulated government project were demanding for a small start-up; to fulfill NASA safety requirements, everything from the heaters that soften the plastic for printing, to the wiring that powered them, to the fumes from the melted plastic, had to be tested and approved by the space agency. Of course, the printer also had to function reliably in zero-g—no small feat in itself. Tests on the ISS were successful, with a number of useful plastic parts manufactured. The company's founders have far larger visions for the future, including printing with metal and even dirt from asteroids, the lunar surface, and Mars. Made In Space is currently designing 3-D printers that can work with metal to process materials found on asteroids into useful products for spaceflight, larger 3-D printers for use in the ISS, and even the manufacturing of specialized fiber optics in space.

Not all space companies will make products that operate *in* space. Kymeta was founded by former Microsoft executive Nathan Myhrvold with funding from Microsoft and other investors; it's designing a laptop-sized unit that will be able to track and communicate with satellites in low Earth orbit. Maintaining a signal with these constantly moving targets can

be difficult, but Kymeta's new system promises to increase reliability at greatly reduced cost. The unique technology can direct a radio beam toward a mobile object in space without the use of moving parts. Metamaterials—special chips that can "tune" the direction of incoming signals—allow Kymeta to use lightweight, flat receivers to accomplish the same tasks that previously required steered parabolic dishes.

There are hundreds of other small enterprises jumping into the commercial space sector with unique high-tech products. Some are well-funded start-ups, while others are being created in university labs and garage workshops. Some have small NASA contracts, but most are operating on their own dime. NASA has long worked with small vendors, but it's diversifying how it does so. The agency supports small start-ups and university students with "hackathons" to promote new software and technology designs. Much of this support takes the form of competitions. One example is the "Space Poop Challenge," which, despite its whimsical name, sought to address a very real problem in spaceflight: the elimination of waste from the human body. Traditionally, special diapers have been used in space suits to contain waste from astronauts during EVAs and long periods strapped in their seats, but extended exposure of skin to urine and feces can create irritation and even infections. The competition was launched to find better solutions and attracted 20,000 competitors from 130 countries, who submitted 5,000 completed designs. The $15,000 grand prize was awarded to Thatcher Cardon, an Air Force flight surgeon from Florida, who prototyped his winning entry with parts bought from thrift and budget stores.[49] Runners-up included students and consumer product designers. Such outreach to small inventors and start-ups is part of the "new NASA," striving to find improved and more cost-effective solutions to the daunting problems of spaceflight.

This is but a small sampling of the ways in which bright individuals and entrepreneurs are embracing the space trade. These are people who, just a decade or two ago, would likely not have entered the business. It's a big move away from the old-school methodologies established during the space race, when money flowed to mostly large, established suppliers. But not all the participants in Space 2.0 are small operations. A few large companies, started by well-heeled internet billionaires, are taking the lead in building the new space economy. One of these luminaries is the Tony Stark of Space 2.0: Elon Musk.

Image credit: James Vaughan

CHAPTER 8

SPACE EXPLORATION TECHNOLOGIES CORP.

Travel about two hundred miles west of the Mojave Desert, and the world turns upside down. Baked desert pavement becomes rolling hills shaded with green, and soon the sparkling Pacific peeks through the valleys ahead. The Central California coast is stunning. If the high desert is defined by flat and dry, the state's western shores are characterized by plunging terrain that ends in a rocky shoreline. On the steep bluffs between sea and sky, the brisk wind can bring a chill, with temperatures in the low fifties. While just a few miles inland the region is dotted with housing developments and strip malls, twenty-two square miles of this stunning landscape remains largely undeveloped. This windswept, scrub-covered terrain is home to Vandenberg Air Force Base, just west of the small town of Lompoc. The base was founded in 1941 and has long been the Western Test Range for America's nuclear missile forces.

There are many relics of the space age at Vandenberg. Mothballed nuclear missile launch sites, remnants of an Air Force space

Shoreline of the California Central Coast near Vandenberg Air Force Base. Image credit: jakobradlgruber/123RF.COM

SpaceX hangar at Vandenberg Air Force Base. Image credit: SpaceX

Main entrance, Vandenberg Air Force Base, California. Image credit: Claudine Van Massenhove/123RF.COM

station program that was cancelled in 1972, and a stillborn launch complex intended to support military space shuttles from the West Coast dot the landscape. But one facility stands out. The paint is a bit whiter, the buildings newer, and the signage fresh. This is Pad 4, known more formally as Space Launch Complex 4E (or SLC-4E), SpaceX's new California facility, where the company launches its Falcon 9 rockets on the West Coast. SpaceX leased SLC-4E from the Air Force in 2015 to accommodate launches and eventual landings of the reusable Falcon 9 first stage, though until 2018, they returned to SpaceX's autonomous recovery barge far out in the Pacific—it took a number of years for the company to gain permission from the state of California to land back at Vandenberg.

A missile launch complex under construction at Vandenberg Air Force Base in 1966. Image credit: US Air Force

A Falcon 9 rocket heads to the launchpad at Cape Canaveral, Florida. Image credit: NASA

It's December 22, 2015, and SpaceX is about to stage its dramatic return to flight after a Falcon 9 rocket exploded in Florida in June. There is a lot riding on this event. One might argue that the credibility of the world's newest and boldest rocket company is at stake.

SpaceX also launches from the Cape Canaveral Air Force Station in Florida, and from NASA's Kennedy Space Center, at Pad 39A. The latter is the historic site from which most of the Apollo spacecraft and the space shuttles launched. But rockets originating from the cape arc eastward out over the Atlantic, headed for orbits around the earth's equator. The unique quality of Vandenberg is that it allows for trajectories that run perpendicular to the equator, inserting satellite payloads into polar orbits, which circle the earth in a north-south orientation. This is an ideal place to launch both military spy satellites and commercial Earth-imaging craft; from this orbit they can look down on the entirety of the planet as it slowly rotates below them. On this day, the Canadian Space Agency's CASSIOPE space weather tracking satellite is perched atop the steaming Falcon 9 rocket.

I joined about twenty attendees from the media atop a rocky bluff to cover this important launch. Had this been a NASA event, we'd have a press representative, folding chairs, speakers wired to launch control, and a flashy countdown timer. Maybe even coffee. SpaceX's spartan control center, down at the other end

A Falcon 9 rocket prepares to launch from SLC-4E at Vandenberg Air Force Base. Image credit: US Air Force

of the base, remained closed to the press, so we would see the launch from this desolate dirt parking area. This would have been fine if we'd been given some way of knowing what was going on, but we had no contact with the control center—no radio, no walkie-talkies, *nothing*. Just a general awareness that there would be a launch this morning off to the south. This event occurred early in SpaceX's West Coast operations, and the finer details of handling the press had not been worked out fully—for now, the Air Force was handling media relations.

Eventually, a couple of young airmen pulled up in a truck, unspooled some wire, and set up a single battered speaker. The spectators now had a launch audio feed—but it was an engineering channel and did not carry the countdown commentary. We learned which valves had been opened and closed, and that tank pressures were holding, but not much more.

A bit more fiddling on their part availed us little, so after about twenty minutes, I started doing video interviews of some of the Air Force people about their work with SpaceX. At about 9:20, I was midway through interviewing Second Lieutenant Kaylee Ausbun, our Air Force public relations officer for the morning, when she said, "We're here to see the launch of the Falcon 9 rocket today. This is a historic event for Vandenberg, as this will be our third launch within a week. The last time this happened was in 2004, when—" She abruptly stopped talking and gawked past my shoulder. There was a collective gasp, and I twisted around to see the rocket in midclimb. The rumble of the launch arrived a few moments later, just about the time I finally got my camera focused on the ascending Falcon 9. My curses were louder than the continuing, uninformative engineering commentary, but I learned a valuable lesson that day: *Always keep your eyes on the rocket.*

A Falcon 9 launches from Vandenberg Air Force Base. Image credit: SpaceX

That's exactly what Elon Musk has done since he started SpaceX: He's kept his eyes on his goal—building and flying ever-larger rockets. His Falcon 1 got him started in spaceflight. The Falcon 9 allowed him to gain paying customers and prove that his rockets could return for reuse. The Falcon Heavy, the largest rocket to fly since the Saturn V, launched in 2018 and will continue to do so with increasing frequency, allowing for record-sized private payloads to be delivered to space.

Musk formed Space Exploration Technologies Corp., better known as SpaceX, in 2002 and the company successfully launched its first Falcon 1 in 2008, after three failed attempts. The Falcon 9 came next in 2010, clustering nine of their successful engines to create the much larger rocket. The Falcon 9 has launched over fifty times, in ever-improved iterations, with a 96 percent success rate as of today—comfortably within industry standards.[50]

In the beginning, Musk invested a hefty portion of his personal fortune into the company. He had grown wealthy via a number of software ventures, most notably PayPal. He has since spent well over $100 million of that money on SpaceX, an investment that has netted him NASA contracts worth billions, as well as launches for the US Air Force and private satellite companies.

Elon Musk, founder of SpaceX. Image credit: SpaceX

Elon Musk meets Charles Bolden, NASA administrator, 2012. Image credit: NASA/Bill Ingalls

SpaceX, like Blue Origin and a select few other companies in Space 2.0, is largely personality-driven. It's a big-vision company, directed by a larger-than-life individual with big ideas.

"If the objective was to achieve the best risk-adjusted return," Musk noted in 2016, "starting a rocket company is insane. But that was not my objective. I had certainly come to the conclusion that if something didn't happen to improve rocket technology we would be stuck on earth forever. And the big aerospace companies had no interest in radical innovation. All they wanted to do was make their old technology slightly better every year, and sometimes it would actually get worse."[51]

Musk set out to redesign the way rockets fly. His innovations range from more efficient fueling procedures to landing and reusing the boosters and nose cones, and eventually the second stages. His

A SpaceX autonomous landing barge, which provides a seagoing landing site for returning SpaceX rockets. Image credit: SpaceX

ultimate goal is total reusability—nothing will be thrown away. It is worth noting that SpaceX's accomplishments owe much to the past achievements of NASA and traditional aerospace companies during the space race, a fact Musk readily acknowledges.

As of 2018, SpaceX is launching rockets from both coasts of the US, owns a small fleet of ships for recovery and transport, and has construction and launch facilities in California, Florida, and Texas. Musk has set a number of goals for SpaceX: build rockets, make them reusable, make them bigger, lower launch costs, and launch frequently. So far he has met most of these milestones, though he still works to ramp up to the launch schedule he covets: rockets leaving Earth daily.

The idea of a private individual starting a company to build and fly rockets is not unique; many had tried to do the same in the 1980s and 1990s. What is unique about Musk and SpaceX is that he succeeded—he is flying commercial and military payloads every month, and generating substantial revenues. So far, Musk has launched cargo to the ISS for NASA, satellites for a number of private and government entities, and military payloads for the US Air Force. This last customer was the toughest to gain. The Air Force had a long-standing contractual agreement with SpaceX's only domestic competitor, United Launch Alliance. This agreement obligated the Air Force to pay almost a billion dollars per year for ULA to remain on standby for high-priority military launches—and ULA got the money whether the launches occurred or not.[52]

Musk wanted some of the Air Force's business and thought that the money ULA was receiving was more of a federal subsidy than a true standby fee. He felt that this was anathema to competition. So in 2014, after repeated attempts to negotiate permission to bid on Air Force launches, he sued them.

This action was not taken out of petulance. As Musk put it in a 2014 interview, he had been trying to crack the US military launch market for years, and, "We were essentially left with the only option, to file a protest." Gwynne Shotwell, SpaceX's president, later added that national security launches were "by far the largest single market" for the company's services, exceeding $3 billion annually.[53]

This legal action was a very public declaration of intent. Musk spoke about it before Congress and in the press. His contention was not only that the billion-dollar subsidy to ULA was unfair and anticompetitive, but that his company could save taxpayers money. He merely wanted SpaceX to have the opportunity to participate in the bidding process. He further pointed out that SpaceX was the only company publishing its launch fees publicly, and that those fees were at least 50 percent cheaper than ULA's were reported to be. SpaceX would charge

A SpaceX Falcon 9 rocket launches a US Air Force payload from Vandenberg Air Force Base. SpaceX owner Elon Musk had to file a lawsuit to allow him to bid on Air Force launches. Image credit: US Air Force

between $90 and $100 million per military launch (which tend to cost more than commercial launches due to their unique requirements for quick booking with high reliability). He contended that ULA was charging as much as $500 million per launch, though ULA claimed it was closer to $200 million. In any case, SpaceX was clearly going to be less expensive.

But cost shouldn't be the only consideration in doing business, ULA pointed out. Reliability is key, and at the time ULA's rockets had a slightly higher percentage of successes than SpaceX's. ULA also had two launch systems—the Atlas and Delta rockets—and if one suffered problems, the other would provide failsafe access for military payloads. These were its rationales for the elevated fees it charged the government, along with the contention that it had a long-term, proven record of launch successes as opposed to upstart SpaceX. Notably, however, ULA still did not reveal its launch fees to the public.

While suing potential customers might not be the best way to win new business, Musk is hardly a traditional aerospace figure. He was competing in a new arena with some very old and entrenched players, and legal action seemed like the only course forward. The suit was settled less than a year later—both companies were now allowed to bid for Air Force contracts—and by February 2015, SpaceX had lofted its first Air Force payload. The

A US Air Force payload launches atop a ULA Atlas V rocket. ULA was the Air Force's primary launch provider for years, until SpaceX became a competitor. Image credit: US Air Force

It's worth noting that in 2016, the National Defense Authorization Act forced the Department of Defense to specify costs of national security space activity, and ULA's launch costs were more clearly revealed. Documents showed that the Air Force expected to be paying upward of $423 million per launch by 2021 if ULA were allowed a launch monopoly, so SpaceX's pricing appears to be a good value for the taxpayer. This is one more example of how new and disruptive companies will continue to change the space business landscape in the next decade.

Musk certainly isn't done with his disruptions. While concerned about making a comfortable profit with SpaceX, he is also driving down costs with considerable investment in his rocket technology. To this end, the Falcon 9 was designed to carry landing legs and has the ability to refire the engines on its first stage to allow for recovery and reuse, a tactic that has been challenging, but has ultimately proved to be successful. The company has already reflown numerous first stages and Dragon capsules, resulting in a substantial cost savings: as much as 30 percent per flight.[54] SpaceX is currently working to make all the components of its rockets reusable. This includes the second stage and the rocket's fairings (the shell-like nose cone that carries the payload, which is itself worth about $5 million). Musk's

lawsuit was a risky gamble that paid off, and it probably foreshadows how some business will be conducted in the new space area. Like any commercial enterprise, there will be rough-and-tumble encounters along with more congenial bidding for contracts—but that's business.

ultimate goal is to launch a rocket, land the first stage, recover the second stage and fairings, inspect them (and repair as needed), fuel it up, and fly again. Musk hopes to lower the time frame for this process to about twenty-four hours—it's the only way he sees to get launch costs down to the point he wants.

As Musk puts it, "To cut the cost of getting to orbit to just $100 per pound, you need to be able to launch multiple times a day, just like an airplane." It's a lofty goal, given that even with SpaceX's aggressive launch pricing, it still costs upwards of $2,500 per pound to reach orbit.[55] "It's got to be complete, so you can't be throwing away a million dollars of expendable hardware every flight either."

There will be no deliberate ditching of spent rocket boosters in the ocean for Musk once his newest designs are flying. These rockets, intended to come online in just a few years, will always return to the launch area.[56] Musk's contention is that nobody in their right mind would scrap a commercial airliner after one flight between Los Angeles and Tokyo, and you cannot do this with rockets and hope to advance aggressively into space.[57]

Reusing rockets is fancy stuff. While flying them is always hard, and doing it

A Falcon 9 rocket vents excess fuel after landing on an autonomous SpaceX barge in the Atlantic. Image credit: SpaceX

Early attempt to land a Falcon 9 booster on a SpaceX autonomous barge. The booster ran low on fuel just as it set down, then tipped and exploded. It took a few tries to get the process right. Image credit: SpaceX

SpaceX headquarters, Hawthorne, California. Image credit: SpaceX

reliably is harder, enabling them to successfully and consistently fly themselves home has been, until recently, nearly impossible.[58] SpaceX's competitors will be challenged to keep up.

Hawthorne, home to SpaceX's headquarters and main plant, is a suburb of Los Angeles. The HQ is situated near Los Angeles International Airport on the former site of aerospace contractor Northrop Grumman Corporation. Hawthorne is a relatively quiet community, with the motto "The City of Good Neighbors," and to the city's delight, one of those neighbors is SpaceX. In 2016 I headed to Hawthorne to talk to SpaceX's president and COO, Gwynne Shotwell, who works directly under Musk. Prior to the interview, I was given a tour of the plant. It was nothing short of amazing.

I've been around a lot of rockets, visited various aerospace assembly facilities, toured most of NASA's larger field centers, seen shuttles being built, looked on as Mars rovers and other robotic probes were assembled and tested, and examined lots of US and Soviet space hardware close up. It's all been remarkable, but nothing quite prepared me for this. I was soon standing in a bona fide, privately owned rocket factory on a truly mammoth scale—the first of its kind anywhere.

It's hard to describe exactly what makes SpaceX's main plant so special. Where Virgin Galactic is a small shop slowly crafting a handful of spacecraft, SpaceX is vast—somewhere around a million square feet in size—with more than 5,000 people in its extended workforce,

SpaceX's rocket factory in Hawthorne, California. Image credit: Photo by Jurvetson on flickr

busily assembling several capsules in various stages of completion, along with dozens of boosters and engines. NASA has sprawling but aging facilities, working with quiet exactitude to assemble mostly one-off spaceflight components in dust-free environments; SpaceX headquarters is one vast interconnected shop, with squadrons of mostly young engineers, machinists, technicians, and office workers laboring side by side to build rockets—lots and lots of rockets. The day I was there, I saw a trio of main engine assemblies for Falcon 9 boosters sitting near the front of the facility, right next to the well-patronized café and bistro that is open to the shop floor. Moving deeper into the complex, teams were assembling the components of many different Falcons. In the center of the main

room stood a row of fuel tanks, the silver sides of which bore oval machining marks. They looked like someone had roughly run a grinder over the sides of the tanks, because that's exactly what had been done.

These racetrack-shaped grinder patterns are indicative of how SpaceX does business. One thing that has made rockets so expensive in the past is a process called mass reduction—the lightening of the rocket. The Atlas and Delta rockets, the bread-and-butter of SpaceX competitor ULA, use a design called an isogrid to create strong but light rocket fuselages. Simply put, after the cylindrical hull is fashioned from thick aluminum alloy, adjoining triangles, just a few inches across, are machined out of the metal skin on the inside of the rocket. Many thousands of these cuts are required to lighten it appreciably. The mass of the metal ground out of each little triangle adds up, yet the remaining structure remains very strong. The process is incredibly time-consuming and expensive, but when you're building rockets for national defense (both the Atlas and Delta were originally nuclear missiles), cost is not really a factor. Now that competition exists in the launch business and rockets are being repurposed for commercial use, cost has to be considered and controlled

When designing the Falcon, SpaceX looked at mass reduction with fresh eyes. Their lead engineers studied how NASA, their contractors, and even the Russians

An example of isogrid machining. This is the standard technique of "mass reduction" for rockets. Part of the fuselage is created, then triangles are machined out to lighten the component. The resulting structure is very strong, but it's an extremely expensive process, since it is done by hand. Image credit: NASA

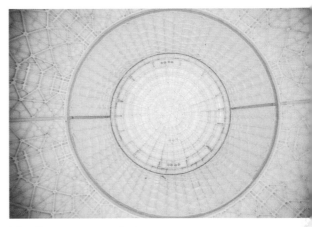

The Dragon capsule is one component on which SpaceX does use isogrid machining. This is an interior view, looking up toward the capsule's nose. Image credit: SpaceX

built their rockets. What they saw surprised them. The construction techniques they utilized were effective but exotic and expensive to accomplish. SpaceX offered a simpler alternative answer to the problem of mass reduction—grind metal off the rocket's fuel tanks in a pattern that will allow them to be lighter but remain very rigid. Though the results are a bit ugly until painted over, the process is simple and it works. It is also vastly less expensive and time-consuming than isogrid machining. This is just one example of how Musk has changed the way things are done in rocket building, with a relentless focus on affordability and reliability.

Deeper into the facility stands a large clean room, a sealed environment with highly filtered air, in which more engine components were being scrutinized and a couple of Dragon 2 capsules were under assembly. Outside, to the left, additional rocket stages stretched off toward the far wall, where engineers constructed parts of the Falcon Heavy.

This is a true rocket factory, building space machines on an industrial scale. While the facility is vast and fairly new,

Casually dressed SpaceX employees look into the glass-enclosed control center at SpaceX headquarters in Hawthorne, California, during a Falcon 9 flight. Image credit: SpaceX

Gwynne Shotwell, president and COO of SpaceX. Image credit: NASA

there are plenty of worn edges—this is a plant focused on volume construction, not the newest and best machinery. Everything is just big enough, new enough, and fancy enough to get the job done, but without excess. Falcon components are lined up in rows, resembling an automotive assembly line. It's a very busy place, and Musk often roams the plant, stopping to inspect a part here or to chat with a small group of employees there. He was dressed casually the day I visited. In fact, there is not a suit or tie in sight. That's not the SpaceX vibe.

I met Shotwell in a small conference room just off the shop floor. Inside was a black table with chips and dents in the surface; it has clearly seen some rough use,

probably from various mechanical and electronic parts being spread out upon it for examination and discussion. At first I mistook it for slate or even Formica, but I soon realized it was just a simple piece of well-used black-painted pine. It's just enough to do the job. No rosewood or walnut here.

Shotwell was born and raised in small communities in Illinois. She originally planned for a career in the automotive industry. After graduating from Northwestern University with degrees in mechanical engineering and applied mathematics, she went to work for the Chrysler corporation, but soon left to move to California. There she spent over ten years at Aerospace Corporation, a think tank for government, military, and, more recently, corporate clients, where she worked on various assignments, including space shuttle operations. Shotwell then spent another four years with a smaller rocket start-up, and in 2002, left to become Musk's seventh hire at SpaceX, as vice president of business development. She's been with SpaceX ever since.

I asked her how the story of SpaceX might read to future grade school students. "It's all about building a company and a capability to ultimately build ships that can take people to other planets," she said. While SpaceX has long focused on Mars as the ultimate objective, Shotwell noted that the moon has also become a part of the company's agenda. Less than nine months after Musk formally outlined his ambitious agenda for a mass human migration to Mars at a major aerospace conference, he announced, "To really get the public fired up, I think we've got to have a base on the moon. Having . . . a moon base, then getting people to Mars and beyond, that is the continuance of the dream of Apollo that I think people are looking for."[59]

This is not really so much a change in corporate trajectories as a smart refocusing of near-term efforts. Mars remains Musk's dream, but the moon will probably be a first stop. Financial realities will impact the choice, but there's more to it than that for Musk and SpaceX.

"There's a business case for Mars at the projected cost for our new system," Shotwell explains. "That said, even if there were no business benefit, I think we'd still do this. This has always been Elon's drive; this is why he founded the company." His vision is to colonize Mars, hopefully at a profit. "Like the East India Trading

SpaceX's business case for Mars: reusable rockets, such as the BFR (Big Falcon Rocket), seen here, transport up to one hundred people at a time to Mars for about $250,000 each. Over time, use of the rockets should pay for their cost—and generate a profit. Image credit: SpaceX

Company of the seventeenth century, if you build it, the money will come. [If] you go and explore . . . you will find your wealth, you will find the business case," Shotwell says. But clearly, if the US government will work in collaboration with SpaceX to send payloads and people to the moon as Musk continues Mars-related research and development, SpaceX will gladly oblige.

SpaceX's view of the future is a grand vision, one involving a lot of new technology and elaborate space infrastructure. To better understand Musk's plans, it might be instructive to take a look at SpaceX's short history.

The Falcon 1 rocket first flew into orbit successfully in 2008—a year that had also seen financial difficulties plague SpaceX. In the wake of a series of launch failures, Musk found that he had burned through the initial funds he had allocated to the company. SpaceX was in dire straits. Musk reached out to an early investor in his Tesla automotive company, Steve Jurvetson, who put together an infusion of cash for SpaceX following an earlier investment of $20 million from a group called Founders Fund.

Continuing to follow his incremental design philosophy, Musk took the now-proven engine from the Falcon 1 and clustered nine of them into a larger, more advanced rocket—the Falcon 9, which first flew in 2010. Various iterations of

A Merlin engine undergoing tests at a SpaceX facility in Texas. Image credit: SpaceX

that rocket have emerged rapidly since that success, each an improvement upon the last. The current rocket, the Falcon 9 Block 5, pushes the rocket's basic design to lift almost double its original specifications, up to 50,300 pounds, to low Earth orbit.

With reusability as a key objective, the company has so far been able to retrieve and reuse the Dragon capsule and booster stage. Recovery of the rocket fairings has proved more elusive, but they have been recovered in tests and will be reused soon. All of these are expensive items and designed to be used many times. The first stage boosts the rocket before releasing the second stage, which separates, ignites, and continues its journey into low Earth orbit or geosynchronous orbit, depending on the customer and its needs. Then, in most flights, the booster reverses direction and fires its engines to slow itself and head home, either back to the launch site or to a robotic drone ship. (The exceptions are flights where all the available fuel is needed for a heavy payload or high orbit, in which case the booster is treated as an expendable launcher, breaking up as it returns to Earth—just like competitors' rockets.) To return in one piece, the booster must slow from almost 3,000 miles per hour to head back toward Earth.

By the time the booster has finished its braking burn, it's gliding along at an altitude of about one hundred miles. It soon reenters the atmosphere, a challenging task for any machine. It's not traveling fast enough to need a heat shield, but the ride home is still a hot and violent one. The engines fire again during reentry to keep the rocket stable and slow it further.

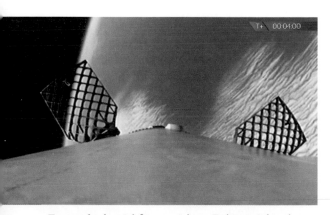

Extended grid fins guide a Falcon 9 back to a landing. Image credit: SpaceX

Once the booster is in denser air, four paddles—SpaceX calls them "hypersonic grid fins"—fold out from its top. The paddles resemble perforated, oversized spatulas and are mounted on pivots to allow them to steer the rocket back to Earth. And here is where it gets really tricky—the rocket must land itself automatically.

Except for those few flights in which the rocket must use all its fuel to deliver its payload, it steers toward a SpaceX landing barge in the ocean, off the coast from the launch site, or lands not far from its launch site on terra firma. In either case, this process is fully autonomous; the computer on the rocket communicates with the landing target and flies itself back to Earth. By the time it has completed reentry and flown to the landing area, the rocket has fired some of its engines four times—once for launch, once to reverse direction, once during reentry, and again for final braking and landing (not all engines are required for the landing phases). Upon landing, the rocket has decelerated from almost 3,000 mph to just 5 mph at touchdown. To date, only SpaceX has accomplished this in practice, although Blue Origin is flying autonomous suborbital test flights that also land autonomously.

In the final moments, the rocket rights itself into a perfectly vertical orientation, and four landing legs, which have been folded up against the fuselage during flight, pivot down and lock, allowing the rocket to land. If this occurs at the launch site, the rocket merely comes to rest on the ground. If it's at sea, on the landing barge, the computers must judge how to best set down, depending on the barge's motion. The landing barge's "sea state," or motions in the sea, can vary by as much as ten feet vertically. To land safely in such conditions is an amazing accomplishment. After a few early failures, both barge and dry land returns have become relatively routine for Space X.

The rocket, capsule, and fairing may then be inspected and refurbished for another use, removing one of the major expenses of launch vehicles—manufacturing. Reuse of these components has already resulted in cost savings, some of which are passed along to the customer.

Eventually, Shotwell says, "We'll be selling Falcon 9s in single-digit millions. That's the ultimate goal—the theoretical cheapest price." Falcon 9 launches at about $60 million now, so a reduction of even that bargain price will be welcomed. The company has said that the rockets and associated hardware are the primary cost; fuel cost is about $200,000. Additional expenses are the hours needed to inspect and refurbish, if necessary. So if SpaceX can recover and reuse the booster, the second stage, the fairings, and sometimes the capsule (satellite launches don't use a Dragon capsule), the potential savings will be huge and may soon approach 50 percent.[60]

The Falcon Heavy is the next step in Musk's march to interplanetary travel. Its maiden flight took place in February 2018 and was a success. The rocket reached orbit, deploying a test payload composed of Musk's personal red Tesla roadster with a space-suited mannequin inside. Both side boosters landed successfully back at Cape Canaveral. The only complication arose when the center booster core missed the landing barge at sea; it crashed into the ocean nearby. In Musk's view, that failure was a developmental detail. The engineering data gathered from the flight gave SpaceX enough information to avoid such mishaps in the future. The crash was caused by insufficient igniter fluid to relight the engines for the final landing burn.

Three Falcon 9s strapped together compose a Falcon Heavy and can lift about triple the payload of a single Falcon 9—up to 140,000 pounds to low Earth orbit. It's the largest American rocket to fly since the Saturn V moon rocket of the 1960s, and the largest in the world since the Soviet Union's Energia booster of the 1980s. The Falcon Heavy will enable large payloads to fly to the moon and possibly beyond. By using three Falcon 9s in a cluster, SpaceX is building on their proven technology. The Falcon Heavy uses twenty-seven of Musk's Merlin rocket engines, instead of the five huge rocket engines employed in the Saturn V.

And what of Musk's Mars plans? In 2016, he announced an audacious design for a huge Mars craft with the capacity to send up to one hundred people to the Red Planet in a single flight. The BFR ("Big Falcon Rocket"), which Musk renamed "Super Heavy" in late 2018, will dwarf even the Falcon Heavy and produce more than twice the thrust of the Saturn V. The spacecraft atop it (the BFS or "Big Falcon Spaceship," which Musk also renamed as "Starship," is a large, crewed structure that has a pressurized (and habitable) volume equivalent to the Airbus A380 jumbo jet. Besides being Mars capable, the Super Heavy/Starship will be employed by SpaceX to fly payloads to the moon. SpaceX is also developing technology to enable the rocket's use as point-to-point transportation on Earth. In this configuration, the spacecraft would be capable of launching from New York, flying suborbitally, then landing in Shanghai, China, all in about forty minutes.[61] Cargo-carrying variants are on the drawing board, as well.

The Super Heavy/Starship is currently under construction a few miles from SpaceX's main factory. The company has leased a large tract of land in Los Angeles's harbor, which will allow it to build the spacecraft there and then transport it to Florida via barge.

How will the company pay for all this? SpaceX has already secured its first contract for the Starship—a Japanese billionaire, Yusaku Maezawa, who will loop the moon with six to eight artists, and an astronaut or two, as his crew. Shotwell thinks that

The Falcon Heavy prepares to launch on its maiden flight on February 6, 2018, from Florida. Image credit: SpaceX

further partnership with NASA is likely. SpaceX already has two such agreements in place—the space agency has been paying it to deliver cargo to the ISS since 2012—and SpaceX will soon fly astronauts there aboard the Dragon 2. She hopes that NASA will pursue more public-private partnerships along the lines of COTS, the Commercial Orbital Transportation Services program; the space agency used COTS to coordinate deliveries to the ISS with the help of private industry.

"I think the COTS program has been the best bang for the taxpayer buck that I am aware of in any field," Shotwell says. "So doing that for Mars would be great. And if NASA can help to offset the costs, in about the same percentage as they did for us with COTS, then even better."

Through a variety of partnership agreements, NASA paid for about half the costs of developing the Dragon capsule, according to SpaceX. It should be noted that the total cost of that program, by NASA's own estimates, was substantially less than it would have been had the agency undertaken such development on their own. An internal NASA study indicated that SpaceX's development cost for the Falcon 9 and Dragon was about up to 90 percent less what it would have cost to accomplish the same results through traditional contracting arrangements.[62] This study did not include savings garnered by recycling rockets and capsules; it was simply looking at the design and development costs.

NASA has invested specifically in the rocket components it is using to resupply the ISS—the Falcon 9 booster and the Dragon capsule. The Falcon Heavy and BFR have been developed, so far, entirely with company funding. This is a very different process than has been followed by traditional aerospace contractors in the past, where NASA and the Air Force footed the bill for the majority of development for rockets and spacecraft, once contracts were signed. SpaceX's funding arrangement is a preview of how such projects will be pursued in the future, with private operators paying part of the bill and NASA paying for fully or at least partially developed hardware.[63]

SpaceX is engaging in one more large program—to provide global broadband access to the internet from space. The company is planning to launch 1,600 small satellites in six years, and a total of 4,425 within a few more years, to enable high-speed broadband access around the world. These satellites will fly in low-altitude orbits of a couple hundred miles, unlike most communication satellites, which hover in geostationary orbits at about 25,200 miles above the Earth's surface. The large number of satellites means that one can seamlessly hand a broadband signal off to another as they cross overhead, making for uninterrupted global reach. While ahead of the pack, SpaceX is not alone in this endeavor. Another company, OneWeb, has been planning a similar system for years, using 650 or more satellites. Others are eyeing this potentially lucrative market as well.

How quickly this plan comes to fruition at SpaceX, and how soon it can turn a profit, has much to do with lowering launch costs. This gives SpaceX an advantage. It's a classic example of vertical integration, where all the needed services can be provided within one company, keeping costs down and increasing profits. The global broadband satellite initiative is already a hotbed of competition, with the various players working to clear government regulatory hurdles. Besides the Federal Aviation Administration, which must grant permission to operate rockets within US territory, the Federal Communications Commission must also approve the plans, since they will involve the use of government-controlled commercial telecommunications bandwidth. But the potential rewards, to both the company and society at large, well outweigh the costs and reach beyond the financial ledger.

As Shotwell put it during our interview, "We're building a global system of satellites because access to the internet and better education is the great economic leveler . . . If there is global access to the internet, everybody has the same opportunities. So if everybody in sub-Saharan

By building all the rocket components in-house, SpaceX is able to control both quality and costs. Seen here is a completed Merlin rocket engine, bound for the first stage of a Falcon 9. Image credit: SpaceX

Africa has inexpensive, fast access to the internet, that's great. And if people around the world have access to talking about their problems on a daily basis, perhaps you can crowdsource better solutions."

It's a broad statement, and a project not without its complications. Providing inexpensive ground stations to less affluent populations will be a challenge, for example, along with more access to electrical power. But it's a grand vision.

Steve Jurvetson, the investor who stepped up to assist SpaceX in its darkest hours in 2008, concurs. The satellite initiative will eventually "bring four billion people online who are not currently able to access the internet, and that's triple the current online population. Think of the educational opportunities, and possible entrepreneurship. The global economy has never seen anything like it. So when you think of what GPS has done for us, for example, this will possibly be the biggest boost to the economy in history, and is at most five to ten years away." He went on to wonder aloud how many Einsteins might be awaiting us in that enormous, currently under-resourced population.

If the short but amazing history of SpaceX is a guide, these projects will likely succeed, but in longer time frames than suggested. The company will be forced to adjust and reassess as it goes along, as will other companies working in Space 2.0.

But Musk's accomplishments have already defied convention. Alan Stern, the chief scientist for the recent New Horizons Pluto mission, might have said it best back in 2008, when he was the associate administrator for science at NASA: Elon Musk "wants to make people a multi-planet species, and he's not going to quit. He'll change the model, or he'll spend more of his own money—he'll do something. He's not in it to build the rockets; that's a means to an end. It's a religion for him."[64]

In the best tradition of American competition, Musk and SpaceX are not alone in these grand endeavors. There are a number of would-be competitors ready to take up the challenges of a new space race, but only two are anywhere close to SpaceX in terms of financial investment or technical ability: Jeff Bezos's Blue Origin and United Launch Alliance, or ULA.

Falcon 9 carrying CASSIOPE preparing for liftoff at Vandenberg Air Force Base, December 2015. Image credit: SpaceX

Image credit: James Vaughan

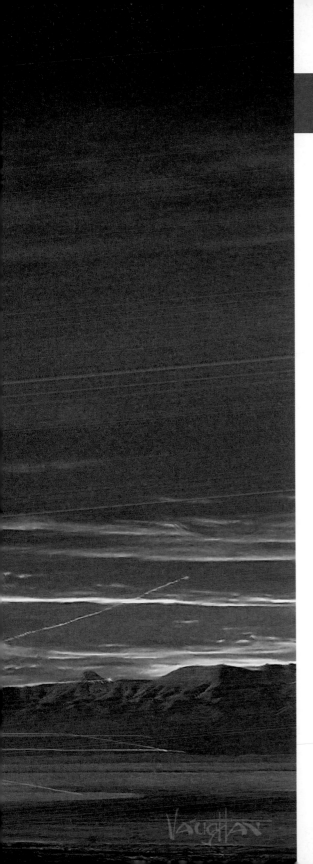

CHAPTER 9

A NEW SPACE RACE

Precious few competitors exist in the human spaceflight launch industry—the resources required to shuttle people into space are staggering. There's SpaceX, as we have already seen, and also United Launch Alliance, which we will detail shortly. But one of the other major players, Jeff Bezos's Blue Origin, truly stands alone. Not because they are doing anything so novel, but because they are doing it so *differently*.

Blue Origin is almost completely self-funded. The company has received a very small amount of money from NASA's Commercial Crew Program—the arrangements that funded both SpaceX and ULA to build their launch systems to ferry astronauts to the ISS—but Blue Origin's share amounts to only about $25 million, a pittance compared to Bezos's personal investment in the company. Subsequently, Blue Origin has submitted a bid to supply large rocket engines to ULA for the Vulcan (ULA's replacement for the Delta IV and Atlas V—more about that just ahead), and won another $45 million contract with NASA to provide suborbital research flights, but it's otherwise operating in financial

135

Blue Origin's planned New Glenn rocket will compete with the Falcon Heavy and should be flying by the early 2020s. Image credit: Blue Origin

Jeff Bezos, founder of Amazon and Blue Origin. Image credit: Blue Origin

semi-stealth mode. The suborbital New Shepard has been entirely self-funded, as is the follow-on and much larger New Glenn heavy booster. It's an impressive and daring path to tread.

Bezos has made his billions with Amazon.com, but he has had his eyes on the stars since his youth. As early as 1982, in his speech as high school valedictorian, Bezos is quoted as saying that he wanted to build "space hotels, amusement parks, and colonies for two million or three million people who would be in orbit. The whole idea is to preserve the earth . . . The goal [is] to be able to evacuate humans. The planet would become a park."[65] Since then, he has developed Amazon into a global marketing behemoth, but space was never far from his mind.

Unlike Elon Musk, Bezos did not cash out of his internet success to pursue building rockets. Bezos started Blue Origin in 2000, just six years after founding Amazon. By all accounts, it has not been an inexpensive undertaking. By mid-2014, he had invested about $500 million into the company, and is now stepping up his commitment.[66] As he put it at a 2017 space conference: "My business model right now . . . is I sell about $1 billion of Amazon stock a year, and I use it to invest in Blue Origin."[67] That's a unique model and reveals an astounding level of belief in his vision. Bezos has sold a total of just under $5 billion worth of Amazon stock since 2004; how much of that went into Blue Origin is not known, but industry insiders estimate it was a significant portion.[68] With an estimated personal net worth north of $150 billion, though, he can follow this course for some time.

Blue Origin has operated quietly for almost twenty years, announcing flight-test successes and the occasional new initiative. Only recently has it released more frequent progress reports and revealed its plans for the future to the public. It therefore comes as a surprise to many that Bezos's rocket company was founded, two years before SpaceX.

Blue Origin is approaching the space business in a very different way than its rivals, SpaceX and ULA. The company currently employs just over 1,500 people and for a long time has focused primarily on building and testing one suborbital rocket, the New Shepard (named after Mercury astronaut Alan Shepard). Blue Origin's initial goal was more like Virgin Galactic's: suborbital tourism. In this case, instead of a rocketplane with wings, it uses a more traditional capsule design on a fully reusable rocket. The New Shepard was first launched in 2015 and has performed flawlessly in eight of nine test flights—the booster was lost prior to landing during its first test flight in 2015, but the capsule did fine, parachuting to a perfect landing.[69] Blue Origin announced plans for a much larger rocket, the New Glenn (named after astronaut John Glenn) in 2016. The New Glenn will be comparable in power to SpaceX's Falcon Heavy, and its first stage will also be reusable.

Blue Origin's New Shepard undergoing a test launch. The rocket and spacecraft have flown nine times, eight of them successfully. Image credit: Blue Origin

Artist's concept of the New Shepard spacecraft that will take six passengers on suborbital excursions. Image credit: Blue Origin

The New Shepard is a single-stage rocket that carries Blue Origin's space capsule on a suborbital trajectory and is designed specifically for space tourism

Blue Origin's BE-4 rocket engine, capable of producing up to 550,000 pounds of thrust. Image credit: Blue Origin

flights. The capsule boasts 530 cubic feet of interior space (or about 300 cubic feet more than the old Apollo capsule) and seats six. It features huge windows all around and is, like the best Space 2.0 tech, reusable. Uncrewed flight testing has been hugely successful. Test flights with human passengers are expected within a year or two.

Blue Origin has also been developing powerful rocket engines for both the suborbital New Shepard and orbital New Glenn. Bezos decided early on to work with an alternative fuel—liquid methane—that is easier to store and handle than liquid hydrogen and cleaner burning than kerosene. The latter is important for reusability, since cleaner engines are easier to refurbish. In addition to the use of these engines on the upcoming New

Glenn, Blue Origin is negotiating to sell its largest rocket engine to ULA for use on the Vulcan's first stage.[70] Of course, ULA is a potential competitor to Blue Origin. This kind of "coopetition," even among rivals, may become a part of the Space 2.0 landscape.

The New Glenn's payload capacity in reusable mode should be somewhere in the same ballpark as the Falcon Heavy's.[71] It's been in development since 2012 and will be flying soon.

I spoke at length to Rob Meyerson, the president of Blue Origin through December 2017, for his thoughts on the company's transition from development and testing to a marketplace presence—moving from speculation to revenue-generating flights. Meyerson earned degrees in aerospace engineering at the University of Michigan and the University of Houston, and worked at NASA from 1985 to 1997 on the space shuttle and a crew rescue system. He then worked with the now-defunct company Kistler Aerospace, which also planned to provide NASA and private launch services with launch vehicles and rocketplanes. While it was not ultimately successful, a number of people, including Meyerson, moved on from Kistler to other NewSpace operations. In 2003 Meyerson joined Blue Origin, at the time merely a team of ten people that would evolve into a company that now numbers well over a thousand employees. Meyerson took a new

Rob Meyerson, former president of Blue Origin. Image credit: Blue Origin

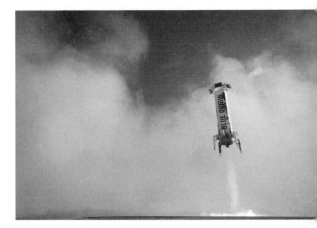

The New Shepard booster comes in for a landing during a test flight. Image credit: Blue Origin

position within the company, running the Advanced Development Programs business unit, shortly after our discussion.

"I see what we all see, a move toward more commercial operations from Earth to low Earth orbit, and that's a really good first step," Meyerson says. "But what we need to do is find that turning point that's going to allow us to increase lifting capacity, and reusability is going to give us an opportunity to reduce cost. That's why we started with New Shepard, our suborbital vehicle. We think that over the next five to ten years, suborbital vehicles like New Shepard are going to offer better access to space, not just for space tourists but also to scientists who have technologies they want to demonstrate in a microgravity environment."[72, 73] According to statements made by Bezos in late 2018, the New Shepard should begin ferrying tourists and researchers on suborbital hops in 2020.

Next we talked about their rocket engines. Blue Origin is developing two large engines: One, the BE-3, has a thrust of 110,000 pounds, and the BE-4 is in the 500,000- to 550,000-pound range. A single BE-3 powers the New Shepard. The New Glenn will use seven BE-4s on its first stage and two BE-3s on the second.

It is quite a daring decision to privately develop a large rocket engine in a marketplace where most of the traditional aerospace engine builders, such as Rocketdyne and Aerojet, have either been bought out or gone out of business. Few of the companies remaining are currently working on new designs, since the US government—big aerospace's favored customer—has not ordered a new engine build for years. SpaceX manufactures its own engines, but these are far smaller than the BE-4. ULA still uses an older rocket engine on the Atlas V that it buys from Russia, but due to the current political climate, Congress forced ULA to look to domestic manufacturers to source its large rocket engines. This was an opportunity for Blue Origin, according to Meyerson,

and the company recently closed a deal to supply ULA with BE-4 rocket engines.

Blue Origin has its eye on other commercial opportunities, some of which will be accomplished in concert with NASA. In 2017, it announced plans to develop a robotic lunar lander largely on its own, which would fly on NASA's SLS booster as well as others. The company has proposed that NASA work to create additional "incentives to the private sector to demonstrate a commercial lunar cargo delivery service."[74] These arrangements would be similar to NASA's contracts with SpaceX and Boeing, in which private industry invests its own money in the design of new spacecraft in exchange for NASA's assurance of use. It's another indication that Bezos is taking his company beyond

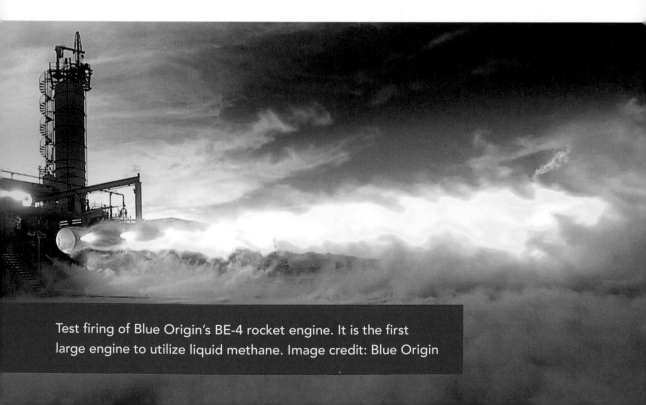

Test firing of Blue Origin's BE-4 rocket engine. It is the first large engine to utilize liquid methane. Image credit: Blue Origin

Artist's concept of the planned Blue Moon lander. Image credit: Blue Origin

internal investment and private partnerships (as he has with his ULA engine deal) and into the realm of government-private arrangements such as those embraced by his competitors.

The projected lunar-cargo delivery spacecraft is called Blue Moon, and according to Meyerson, will "cost-effectively soft-land large amounts of mass onto the lunar surface."[75] Such a spacecraft would fill a huge gap in NASA's current plans to return to the moon—at this time, the government is not funding a lunar lander. NASA worked on such a lander in the early 2000s, but that development ceased in 2009 when the Constellation program was cancelled. Other, smaller companies are working on robotic landers, but nothing approaching the scale of Blue Moon. A privately built and operated lander would provide assured access to the lunar surface for NASA, commercial entities, and possibly international space programs. This would ultimately enable the construction of lunar bases and other infrastructure. Blue Origin is aiming to test the Blue Moon in 2020. It will have a cargo capacity of 10,000 pounds—close to that of the Apollo Lunar Module of the 1960s, but at a tiny fraction of the cost. It will, of course, be reusable.

Not all the companies providing competition for SpaceX in Space 2.0 are like Blue Origin—the recent creations of internet billionaires. In fact, one is an old-school aerospace outfit, mentioned throughout this book, a partnership of two firms that have been in business since the early days of aviation—Lockheed Martin and Boeing. United Launch Alliance, or ULA, has been in operation since 2006.

ULA continues a legacy of American rocketry that goes back to the Cold War.

United Launch Alliance's headquarters in Centennial, Colorado. Image credit: Wikimedia Commons/Jeffrey Beall

ULA's rocketry heritage dates back to the original Atlas, one of America's first nuclear missiles. Seen here are a number of first-generation Atlas missiles being assembled in a Convair plant; Convair was later acquired by Lockheed Martin, which is part of ULA. Image credit: NASA

It currently fields two rockets, the Atlas V and the Delta IV. The Atlas was one of America's earliest rockets, designed to lob nuclear warheads over the Soviet Union, and it gained fame when it was used to boost John Glenn into orbit in 1962. While it was removed from military service in 1965, the Atlas has soldiered on for decades as a satellite booster, a role it continues to fill as the Atlas V via a comprehensive redesign in the early 2000s. The Delta family of launch vehicles began service in 1960, also as a military missile. It, too, has undergone updates and revisions,

though not as thoroughly as the Atlas. Still, both rockets have been substantially modified over the years in an attempt to make them more economical to build and launch. The Delta IV Heavy is currently the second largest, most capable rocket in the US after the Falcon Heavy, but it carries a per-launch price of over $400 million to the Falcon Heavy's $90 to $150 million price (depending on configuration), and is soon to be retired.[76] If the Falcon Heavy enters regular service as planned, it will be a strong contender for heavy payloads.

As of 2018, ULA finds itself in the unenviable position of having been unseated by SpaceX as the exclusive provider of commercial launch services for heavy payloads in the US. The loss of ULA's exclusive agreement with the US Air Force was unsettling. ULA has a very high success rate, slightly higher than SpaceX's, but its rockets also have had more than six hundred combined launches to evolve and mature—SpaceX's Falcon 9 has had about a tenth of that number.

ULA still provides about half of the military and commercial launches in the US and has no intentions of ceding its share of the market to SpaceX or anyone else. It's currently designing a new rocket, the Vulcan, to replace the Delta IV and Atlas V. The Vulcan is intended to be at least partially reusable. It will also be less expensive to build and operate than ULA's current fleet.

ULA has long been, and continues to be, a primary launch provider for the US Air Force. Seen here is a modern Atlas V rocket preparing for launch. Image credit: US Air Force

Due to the traditional methods by which ULA's current rockets are constructed, it's more expensive to launch an Atlas V than a Falcon 9. Until the Vulcan is ready to fly, changes must be made to the Atlas to remain competitive. According to Tory Bruno, president and CEO of ULA, improvements have already been made over practices of the recent past. "It now takes us only half as long to build a rocket in our factory and only about a third as long to assemble it," he notes.[77] The company is not standing pat. They are phasing out the Delta, and the cost of the Atlas has already been reduced by about a third.

Reusability is a big part of SpaceX's operational plan to cut costs, and ULA

United Launch Alliance's Tory Bruno. Image credit: ULA

must work toward the same goal to remain competitive. ULA's Atlas will never be reusable; it's just not a part of the rocket's DNA. But the company is approaching the Vulcan as a clean-sheet design, and its partial reusability will be accomplished following a different path than SpaceX's fleet.

Tory Bruno has been the president and CEO of United Launch since 2014. Prior to becoming the head of ULA, he spent

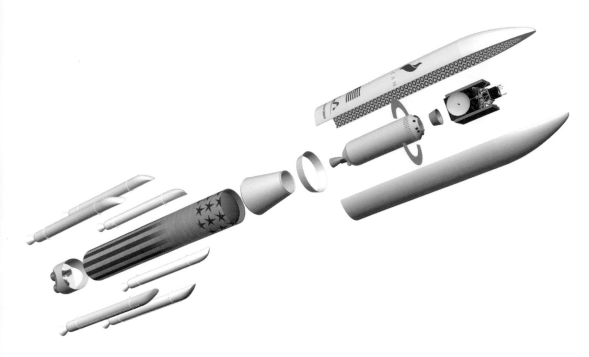

Artist's concept of ULA's new rocket for the 2020s, the Vulcan. It will be partially reusable. Image credit: ULA

thirty years climbing the ranks at Lockheed Martin. Bruno is what is referred to in the trade as a "steely-eyed missile man," having devoted his adult life to the rocket industry. In fact, he's been a spaceflight dreamer since his childhood—he built his first rocket from scratch when he was twelve years old, after discovering an eighty-year-old case of dynamite in his grandmother's barn. I spoke to him from his office in Colorado about ULA's future plans.

ULA has two components to their reusability strategy. The first is to save the booster engines—the most expensive part—rather than the entire first stage. "If you look at the booster, all the money is in the engines," Bruno says. "We have a concept that severs the engines and returns them to Earth in a way that doesn't require any reserve propellant."

SpaceX's rockets must conserve between 15 and 30 percent of their fuel to return to the launch site. This represents fuel that could be used to launch heavier payloads. ULA's plan is to simply drop the engines from the back of the rocket and let the empty fuel tank break up during reentry, like most of today's designs do. The engines will glide back on a parafoil—a parachute that looks like a hang glider—and then be grabbed out of the sky.

Advanced Cryogenic Evolved Stage

75%
PERFORMANCE
INCREASE

ULTRA LIGHTWEIGHT
BALLOON TANKS
3X PROPELLANT

1-4
ADVANCED ROCKET ENGINES
(BE-3U/RL10/XCOR)

Sports Car
to the Stars

SUPER HIGH-EFFICIENT
INTEGRATED VEHICLE FLUIDS
SYSTEM

- ADVANCED INTERNAL
COMBUSTION ENGINE
BURNING WASTE GO_2/GH_2
- REPRESSURIZE TANKS
- GENERATES POWER
- PROVIDES ATTITUDE
CONTROL THRUST
- UNLIMITED ENGINE STARTS

1 SYSTEM
REPLACING **3**

COMPLEX
ON-ORBIT
OPERATIONS

MULTIPLE
ORBITAL PLANES

ULTRA-LONG DURATION

HOURS ➡ WEEKS

ULA's new ACES stage will provide a long-term restartable "space tug" in orbit, reducing the cost of moving payloads to higher orbits around Earth, and potentially out to the moon and back. Image credit: ULA

"We will capture the engines with a big helicopter. That's the way we used to get film canisters back from spy satellites in the '60s, so that's an existing technology and it's a much lower economic hurdle. You can break even in about two reuses and start making money in three," Bruno says.

The second part of ULA's reusability strategy concerns the rocket's upper stage. Traditionally, upper stages have also been disposable items, a situation that both ULA and SpaceX want to change. SpaceX plans to fly back the second stage to be recovered. ULA plans to leave theirs in space.

"We call this new upper stage ACES, which will be reusable in space," Bruno says. ACES stands for Advanced Cryogenic Evolved Stage. He feels that there's not been enough thought given in the industry to reusing a rocket's second stage once it is in orbit. "You use a bunch of energy to get it into space, and you have to take all that energy out to bring it back down. Our first stage is suborbital; it's going to fall back to the Earth . . . that is a much easier job than getting your upper stage de-orbited."

Bruno explains that ACES stages "are inherently refuelable. They will operate for very long durations, which is another

The Delta IV Heavy is ULA's largest rocket and was the largest in the US inventory until the premiere of SpaceX's Falcon Heavy. Unfortunately, it's about five times as expensive to build and launch, and will soon be phased out. Image credit: US Air Force/Carlton Bailie

revolution that, independent of their reusability, changes the things you can do. We will build up a fleet of these space pickup trucks in orbit, waiting to go service spacecraft, or move spacecraft, or swoop down to low Earth orbit to pick up a spacecraft and take it somewhere."

This is an exciting notion, and ACES has generated quite a buzz in the space community. It is, in broad terms, a one-size-fits-all "space tug" that can be parked in orbit for months, possibly years; be refueled as needed; and then directed to accomplish whatever you want in low Earth orbit and beyond.

"The ACES is a concept [that ULA has] been working on for quite a few years," said George Sowers, until recently a vice president of Advanced Programs at ULA and now a professor at the Colorado School of Mines. "It is a LOX/hydrogen upper stage, high-energy, a balloon, pressurize-stabilized tank, thin-walled stainless steel, a very efficient design."[78]

Bruno notes that getting the rocket into orbit is more than half the battle, so it makes sense to leave useful hardware there. "Once you're in low Earth orbit, you can go anywhere in the solar system, which completely changes what you can do," he says. "I can envision a point in time where people no longer go from the surface in one shot to any destination beyond low Earth orbit—you will always stop there, have an ACES swoop down and pick you up, and it will then take you the rest of the way."

The key to making this work is designing the ACES in such a way that it can be fired, occasionally refueled, and refired once in space. This is harder to accomplish than it sounds—rocket engines are generally serviced by technicians on the ground in between uses, and they can get finicky after being fired even once. And while in-orbit refueling has been accomplished before, it's a complicated process with countless possibilities for failure.

An Atlas V rocket being moved prior to launch. Image credit: NASA/Glenn Benson

Regarding fuel, the most powerful ones are stored cold, in what is called a cryogenic state. This is fine over the short term, such as prior to a launch, when they can be topped off as temperatures rise and fuels boil off. But once in orbit, as the cold fuel warms, it expands and is vented into space. ULA plans to counteract this with innovative insulation techniques—basically creating an oversized thermos bottle in which to store the fuel while the ACES awaits its next task in orbit. ACES will also recover the venting fuel as it boils off. "It is basically a highly insulated jerry can that we bolt to the top of the stage, that will allow the ACES to run upwards of one to five years," Bruno notes.

This will greatly broaden the rocket's longevity and usefulness. If customers know there are space tugs awaiting their payloads, ready to move them around in Earth orbit, boost them to higher orbits, or even send them deeper into space, they will not need to spend so much to launch heavy and expensive upper stages from the ground. It's an ingenious concept with a great deal of promise.

As we have seen, much of the engineering of these twenty-first-century rocket builders is aimed at lowering costs.

If launch prices for large payloads eventually fall into the $10 to $20 million range, as has been predicted by Musk, demand could soar, supporting a robust and diverse group of launch providers. Of course, the Europeans, Russians, Indians, and Chinese will not sit idly by and watch US companies dominate the launch market. Each of these nations have built and flown rockets for years and have become good at it, and they are now looking at ways to lower their own launch costs in order to be more competitive. But the US companies are at the forefront of reusability and cost reduction, with SpaceX currently in a comfortable lead, though Blue Origin and ULA are doing their best to keep things interesting.

The plans future of ULA, Blue Origin, and SpaceX are inspiring and bold, with the latter two even more impressive in that they are so heavily funded by single individuals. But most companies gearing up for private and private-public space operations do not have access to the levels of cash that Bezos and Musk do. They will require outside investment to reach their goals. Fortunately for them, there are many investors, both large and small, eager to finance the people creating Space 2.0 and expanding its horizons.

The business end of an Atlas V rocket, with two Russian-built RD-180 rocket engines visible. ULA is phasing out its use of Russian components. Image credit: NASA

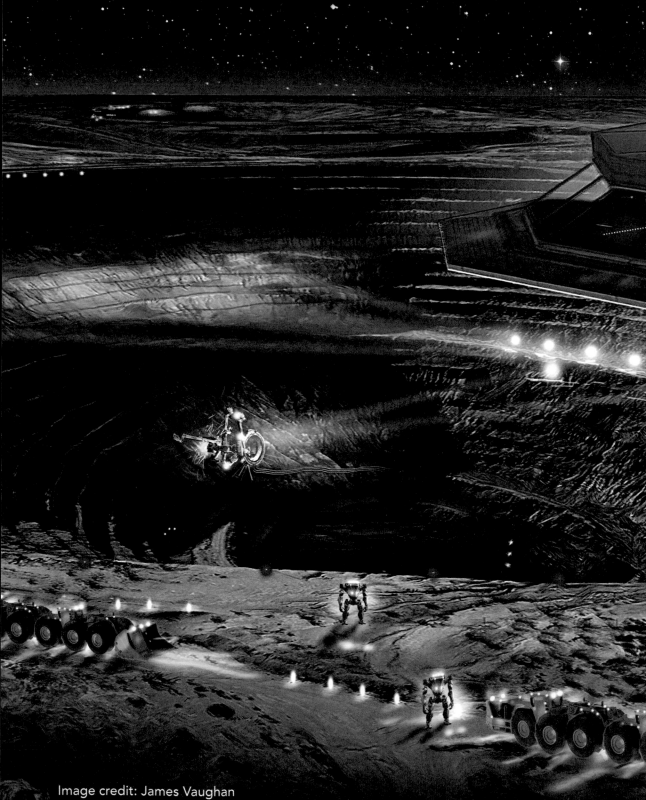

Image credit: James Vaughan

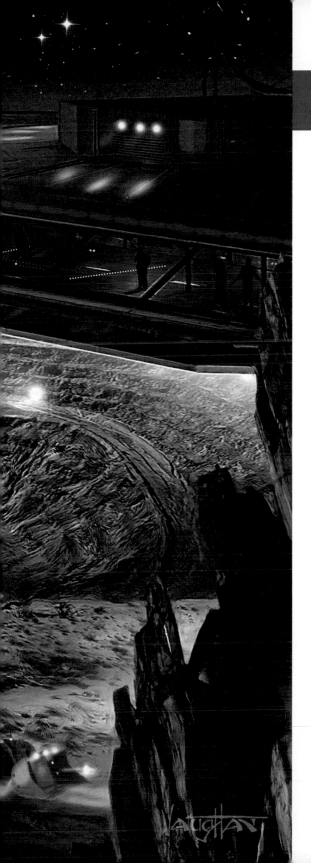

CHAPTER 10

INVESTING IN SPACE

Each year, *Forbes* magazine creates a ranking of the world's billionaires. For 2018, there were 2,208 people on it, with 585 of them from the United States. If we could simply encourage each of these folks to start building their own space program (excluding Jeff Bezos, who is currently at the number-one position and is already doing so), we would see commercial space development move much faster. But Bezos, Musk, and a few more like Paul Allen aside, individuals in the billionaires' club won't be the ones financing the majority of private space efforts. In the coming decade, if current space investment trends continue, the money for space program finanacing will come from larger pools of midlevel individuals, from regular folks who now invest much of their savings in mutual funds. Increasingly, people who are able to invest from tens of thousands to the low millions have come to recognize the potential for both long-term and, eventually, short-term gain from investing in space.[79]

Money must be part of the conversation when we discuss Space 2.0. The days of big government underwriting all space efforts, at least in the US, ended in the 1970s. Government money will continue to play a huge role—via direct investment in programs, the hiring of contractors to supply those programs, and partnerships with private providers such as SpaceX, Blue Origin, and others. But the big future driver for Space 2.0 lies in steak-and-potatoes capitalism and its ability to release entrepreneurial energies on a national, then global scale.

The idea of broad-based private investment in space is not entirely new. Privately financed satellite communications businesses have produced annual revenues on the order of one hundred billion dollars and more since 1980, and satellite-related concerns have prospered since that time. The companies creating the hardware and software that power these enterprises have flourished. Many of these companies are publicly held, funded by portfolios that contain investment from millions of largely middle-class people.

But this is the satellite industry, which has been around since the 1960s. When discussing the early days of NewSpace, what comes to mind are the first companies that attempted to launch their own rockets. They were mostly small, underfinanced operations, and they were not in a position to source funding in the same way as the satellite and telecommunications industry. Let's look at some early

examples of space start-ups and how they have led to the current marketplace for private investment in space.

In 1980, a private rocket company called Space Services Incorporated (SSI) built and launched a small test rocket that flew to almost two hundred miles altitude successfully, well beyond the edge of space. Unfortunately, SSI was undercapitalized and only managed a few more test launches, most of which failed. Since then, the company has been sold a few times. Today SSI operates successfully as a broker for launch services on other companies' rockets. Still, it has the distinction of being an early player in privately funded rocket development.

Private launch company Orbital Sciences was founded in 1982 and in 1990 started flying the Pegasus, the world's first successful privately financed launch vehicle. The Pegasus is an air-launched rocket capable of lifting almost 1,000 pounds to orbit. Orbital Sciences merged with aerospace company ATK in 2014, and the resulting operation continued with the Pegasus. Orbital ATK was in turn acquired by aerospace giant Northrop Grumman in 2018 and rebranded Northrop Grumman Innovation Systems. Innovation Systems continues to supply a myriad of defense and aerospace products, including the solid rocket boosters for NASA's upcoming SLS rocket, and will continue to provide cargo flights to the ISS by the Cygnus

Orbital ATK's Antares rocket, which delivers cargo to the International Space Station under contract with NASA. Image credit: Northrop Grumman Innovation Systems

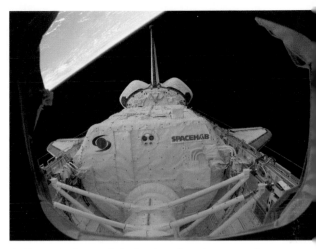

A Spacehab module in the shuttle cargo bay. Image credit: NASA

spacecraft, carried on the Atlas V and its own Antares rocket. Besides the Pegasus and Cygnus, the company has developed a long series of rockets and rocket engines for the commercial market.

In 1984, a company called Spacehab was established to design and build space habitats for use in the shuttle's cargo bay. Its pressurized modules enhanced the orbiter's microgravity experimentation equipment and abilities and expanded its capabilities as a research platform. Spacehab's equipment flew on twenty-two shuttle missions and, in its current iteration as Astrotech, the company has flown hardware aboard the ISS. Spacehab was one of the brighter success stories of the early NewSpace era, with hardware reaching space via an arrangement with NASA that resembles some of what we see with Space

2.0 companies today. It was one of the earliest public-private space partnerships of its kind.

Sea Launch entered the private space market in 1995. While the company was a somewhat predictable consortium of traditional aerospace concerns from the United States, Russia, Norway, and Ukraine, its self-finanacing was innovative for the time (though the connective tissue between the Russian and Ukrainian partners and their governments is more complex). Their Russian-Ukrainian–built rockets were assembled on a floating barge based in Long Beach, California, then transported to equatorial regions for launch. These boosters were capable of carrying 13,000 pounds of payload to orbit. After a total of thirty-one launches and a number of financial challenges, as well as a stint in bankruptcy, the company launched its last payload in 2014.

These are just a few of the notable early efforts in nontraditional launch services built with private funds. They found success in various ways: Orbital ATK, now part of Northrop Grumman, continues to thrive. Sea Launch operated for almost two decades, while Spacehab evolved into a provider of other space-related services. Space Services Inc. survived, though not as a rocket builder. Many other smaller companies tried to develop privately financed launch vehicles but failed for one reason or another. In sum, the time was not right, the market not sufficiently robust, and the money available to start a rocket company from scratch insufficient.

This changed around 2000, when entrepreneurial spaceflight began to develop in earnest. The most obvious new players here were the "billionaires' club" of Musk and Bezos. But smaller entities such as NanoRacks also started raising funds from outside investors to pay for spaceflight operations. Since 2010, investors have provided well over $14 billion for private space efforts, which does not include the billionaires who are investing in their own tech.[80] In 2015 alone, more private space investment was procured than in the fifteen previous years, and this rate of investment continues to climb. Seven companies have attracted over $200 million each.[81] Venture capitalists and investment funds that see the potential of investing in space are driving the trend. A 2017 report cited $3.9 billion invested in Space 2.0 ventures in that year alone (72 percent of which was invested into commercial launch services).[82]

For each space-oriented investment group, there are thousands of dreamers all over the world waiting for their chance to contribute to the opening of the final frontier, but without their own vast reserves of wealth to do so. As you read these words, hundreds of private space-related companies are preparing to enter the marketplace, with more forming every month. These range from small launch providers to tiny satellite builders, from modest bootstrapping operations to well-funded major ventures. This market expansion—a new sort of space race, with competition between boardrooms, not countries—is expected to continue for the foreseeable future.

There are a number of ways to participate financially in the private space sector. For small investors, the safest route is through mutual funds that include aerospace as a prominent part of their portfolio. This is best for people who want in on the potential profits without having to extensively research the marketplace, and those who want to minimize risk. Other investment groups research and concentrate specifically on space-related investments in both established aerospace companies and newer companies—including promising start-ups.

Among the large investors you'll find people such as Steve Jurvetson, a former partner at Draper Fisher Jurvetson Venture Capital (DFJ), who has been responsible for billions in high-tech placements, much of this in space-related companies. While major investors can be found around the world, the vast majority of investment currently originates from the United States. The American business environment is unique and most of the large space-related commercial space enterprises are located there. But foreign investment is on the rise. In 2017, for example, Virgin Galactic announced a deal to receive a billion dollars from Saudi Arabia, adding to the hundreds of millions it has already raised in the Middle East.[83]

Chad Anderson serves as CEO of Space Angels, which handles investments ranging from the hundreds of thousands to single-digit millions. The firm was started in 2007 to cater to a small group of passionate, moneyed individuals who wanted to channel funds into the beginnings of Space 2.0. The Space Angels network headquarters is located in New York with satellite offices in San Francisco, Los Angeles, Seattle, Stockholm, London, Zurich, and Hong Kong.

I spoke to Anderson to get an idea of the range of investment opportunities

Chad Anderson, CEO of Space Angels. Image credit: Space Angels

available, who is investing, and what kind of companies Space Angels found attractive.[84] Anderson gave me a short history of his company.

"From 2007 to 2012, we operated like a traditional angel group, where we played the role of connector. In the early days there wasn't a whole lot of entrepreneurial activity in space. At that point it

The Falcon 1 launches successfully in 2008.
Image credit: SpaceX

NASA-sponsored events, such as this one in Denmark, provide opportunities for students to evolve their ideas into fundable start-ups in Space 2.0. Image credit: NASA

was primarily a credibility-building exercise." Anderson felt that Space Angels needed to prove to the marketplace that investing in space ventures could be profitable.

Each of the four founding members of Space Angels had previously put money into a space technology company. After the first successful launches of SpaceX's Falcon 1, followed by the Falcon 9 in 2010, the partners realized that the private space sector was heating up.

"Things really started to come together at that point . . . the number of business plans that we saw were increasing, and the quality of deals that we saw was increasing." Space Angels now has over two hundred investors placing tens of millions of dollars into small companies. Anderson feels that the industry is poised for exponential growth.

Space Angels has coupled traditional investment thinking with a passion for a new industry—private space enterprise. The risks are high and most of the companies that solicit investment are young, and their placements have required an investor who believes in the promise of space.

"We are looking to back up entrepreneurs who are making a significant impact on humankind, and the financial returns are a happy by-product of that. Today we've got hundreds of companies that have raised money and are executing their business plans. There is going to be some shakeout, and there are going to be some failures," Anderson says, but he points out that in today's environment, this means a few failures within hundreds of companies, as opposed to the handful of such enterprises that existed ten years ago.

Unlike the company's founders, the newer investors working with Space Angels may not know a lot about spaceflight but like the idea enough to risk their money. Anderson says that many of their newer investors "see SpaceX launches on Bloomberg, and they happen to be investors in other companies, and this piques their interest in the space industry. So they come to us."

To ensure the best chance for financial gains, investment firms like Space Angels need to identify companies that will thrive in this new marketplace, and that's a challenge. "It's all new," Anderson says. "There's not a lot of information available, because this has never happened before. It's still difficult to determine what's real and what's snake oil." But this difference becomes more obvious every day, he adds.

Those ambiguities can make the difference between a success story like Nano-Racks or a quick burnout like dozens of other small enterprises. As with any other business landscape, there are safe bets with moderate returns and risky speculative ventures with huge potential—but identifying the winners is tough. In the 1980s, you could have invested in IBM—a safe bet with reasonable gains—or Apple Computer, a wild-eyed start-up with high potential returns. Both were good companies, but Apple made many investors extremely wealthy. The trick, now as then, is to peer into the future.

"At the moment there are a lot of opportunities in satellites, so a lot of people are developing satellite companies, resulting in a bottleneck in launch access," Anderson says. This burst of mostly small-sat activity has created a demand for smaller, lower-cost rockets, and with this, opportunity for both of these market sectors. He cited these as prime investment opportunities in the present, based on demand that should materialize soon.

Thus far, my discussion with Anderson had focused on unmanned spaceflight. Launching humans into space is much more difficult. The technology is more complex, and the needs for protection and life support more demanding. But with SpaceX and others leading the way, it is clear that space entrepreneurs

With companies such as Boeing entering the private spaceflight arena, investment opportunities have widened dramatically in the past few years. Image credit: James Vaughan

are capable of addressing these complex challenges.

"There's a lot of interest in the investment community in human spaceflight given all the activity in low Earth orbit that NASA is turning over to the commercial sector," Anderson says, referring to NASA's contracts with SpaceX and Orbital Sciences to resupply the space station, and the upcoming commercial crew flights by SpaceX and Boeing.

Space Angels is often the first investor in new companies, and sometimes the first in subsequent rounds of funding. One example Anderson cited was Planetary Resources, a company founded in 2009 to develop asteroid mining via robotic spacecraft. Space Angels has invested in the company at each funding opportunity. "We are a major supporter of their team and what they are trying to accomplish,"

In the future, asteroids may be herded closer to the Earth or the moon by private space mining companies for resource extraction. Image credit: James Vaughan

Anderson says. "Personally, I've been most excited about the water, and by extension the fuel and life support basics, that we'll extract from asteroids. We spend so much money and energy to get to orbit, and once there, you are halfway to anywhere in the solar system."

The potential for mining asteroids has been understood for years, but only recently have credible people come forward with a formal business plan for achieving it. Planetary Resources was an early leader in the field and has received funding from both Anderson and big-money investor Steve Jurvetson, when he was at DFJ.

Space Angels represents early money for a number of Space 2.0 companies. But for more mature companies, such as SpaceX, larger amounts of investment are needed.

Jurvetson is on the high end of the private space investment spectrum. In 2008 he arranged a large cash infusion to SpaceX when it was struggling and Musk was tapped out, having spent over $100 million of his own money trying to successfully launch the Falcon 1 rocket, which had failed multiple times. Jurvetson brought in many tens of millions of dollars of outside investment at this critical time. It made all the difference, and the results for SpaceX have been dramatic, with the company now responsible for over half the launches in the US.[85]

Steve Jurvetson in one of his favorite places—the interior of a SpaceX Dragon 2 capsule. Image credit: Photo by Jurvetson on flickr

Jurvetson was born in 1967 and is just old enough to remember the tail end of the space race. He is truly enamored of space exploration—and entrepreneurs. "After twenty years of doing this, I can say that I love investing in passion-driven businesses," he says. [86] His interest in companies like SpaceX; Tesla, Inc. (also owned by Musk); and Planet Labs (a private satellite imaging company) stems largely from the people who started them. "The founders do not put profits as a number-one priority; they are passion driven. I try to find companies like this when they're just start-ups, just people with a dream . . . so we look for the kind of companies that can change the world, and, dare I say, may change other worlds."

Jurvetson was ready to invest in spaceflight in the early 2000s, but he could not find a suitable company. Having met with entrepreneurs who were touting everything from railgun launchers to space tourism, he simply did not see a business model that made sense to him . . . until SpaceX.

"Prior to SpaceX, not a single company, in my judgment, warranted bringing to my partnership," he says. The companies he reviewed either wanted too much investment or were aiming for an unpredictable part of the market. "An example at the time was fuel depots in space. They make perfect sense, but not prior to SpaceX. Who would be the customer? That was not a leap of faith that I was willing to make.

SpaceX technicians prepare a Falcon 1 rocket for launch in 2007. The company was almost out of funds as it struggled to launch the rocket in 2008. Image credit: SpaceX

"But SpaceX was different altogether," he continues, "though when we first invested they'd had three Falcon 1 failures in a row. I didn't realize at the time that they didn't think they could survive a fourth failure. We invested just before that fourth flight," he says with a laugh.

Jurvetson invested in Planet Labs after meeting an engineer who had moved from NASA to Google to work on cubesats. "I met a guy from NASA who was flying smartphones in rockets to see if they could survive the g-loading and so forth. They had already tested them in vacuum chambers and in other ways. Their argument was that they could make a cubesat and fly it close to Earth and get images as good as others that fly farther away."

Since Planet Labs built its business on inexpensive cubesats, using mostly off-the-shelf electronics and flying in low Earth orbit, its costs were far lower than traditional operators. And as these cubesats fail, their replacements are small enough that they can be piggybacked on other companies' launches.

"Replacements can be launched for a fraction of the cost of larger, higher-orbit satellites," Jurvetson says. "I met with them a few months before we invested and they were working in a garage. Planet Labs has now launched more satellites than everyone else combined."

Jurvetson is interested in working with people like himself, leaders in finance who have the passion, drive, and vision to steer ever-larger pools of investors into space endeavors. These investments will not be restricted to rockets and the other large-scale technologies we are used to seeing. Increasingly, private capital will be backing companies that can prosper with smaller advances—nanosats are one example, materials science is another. As

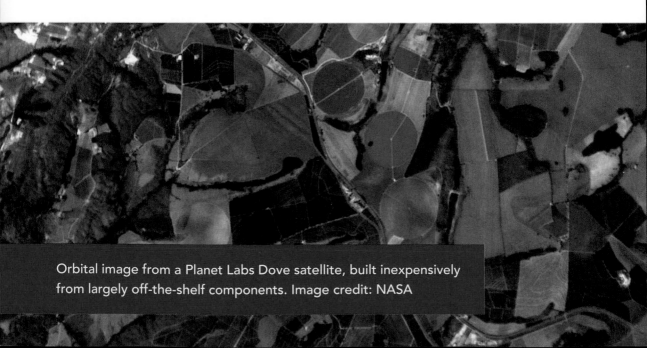

Orbital image from a Planet Labs Dove satellite, built inexpensively from largely off-the-shelf components. Image credit: NASA

financial analyst Shahin Farshchi put it at the 2018 Space Tech Summit, "The next massively successful space company probably won't even be founded, or coined, as a space company, but will derive value by virtue of creating opportunity and capitalizing on the opportunities in space. Someone will come up with a great business where space will be an important piece of that."[87]

Individuals are investing in Space 2.0 in record numbers and amounts. With the rapid growth of the commercial space sector—increasingly fueled by private investment—we will soon have more launch companies, with better associated hardware and services, and more affordable rockets that will be capable of taking people into Earth orbit, to the moon, and eventually Mars.

Up until now, this investment in Space 2.0 has been largely an American activity, though internationally, everyone from tiny Luxembourg to China seems to be stepping up their space sector outlay. While these overseas investors look toward the Space 2.0 players in the US with great interest, there is sometimes significant activity occurring within their own borders. While generally not as entrepreneurial as the US market, the international space sector is thriving.

Image credit: James Vaughan

CHAPTER 11

THE SPACE
BETWEEN NATIONS

While competition is a powerful driver of accomplishment, great things are often achieved through cooperation. We see this every day in business, manufacturing, military operations, and even in politics. In space, a prime example of the strength of international cooperation sails overhead every ninety minutes—the International Space Station. But such cooperation in space was long in coming. When the United States and the Soviet Union decided to enter space in the late 1950s, both nations went solo. Examining the origins of international cooperation in space will help us to better understand the complexities of future efforts.

The space race was an example of extreme nationalism—the antithesis of international cooperation. First, the Soviet Union launched satellites on their own rockets and orbited a man. Then America prepared to send humans on the moon. Since the rockets used to power those early flights were lightly adapted nuclear missiles, both space programs were based on each country's growing ability to wage a

discussions about cooperating in space endeavors. In President Kennedy's inaugural address, he suggested this, saying, "Let both sides seek to invoke the wonders of science instead of its terrors. Together, let us explore the stars."[88] But Soviet leadership displayed little interest. Why is not entirely clear—it may be that they were overly confident, given previous successes, or it may have been that they felt the offer was insincere, a way for America to look good on the world stage.[89]

Soviet leader Nikita Khrushchev wrote a letter to Kennedy in 1962, reopening the subject. This resulted in a limited

In both the US and USSR, early orbital spacecraft were carried by repurposed nuclear missiles. Seen here is the Atlas rocket, carrying John Glenn in Friendship 7 on his historic orbital mission. Image credit: NASA

planet-wrecking nuclear war, so working together was never a strong consideration. But it could have been.

At one of the hottest periods of the space race, as each nation was striving to put a human in orbit and incurring vast expenses in the process, statesmen from both countries sent feelers to promote

President Kennedy and Soviet leader Khrushchev communicated at least twice about cooperation in space, but little came of it until much later. Image credit: The National Archives

Signing of the Outer Space Treaty in 1967.
Image credit: NASA

exchange of scientific information. Then, in 1963, Kennedy raised the possibility of cooperation on a lunar landing, first with Soviet representatives and later that same year in an address to the UN, but little came of it.[90] After Kennedy's assassination, and perhaps in part because of it, the overtures seem to have been forgotten.[91] The race to the moon continued in an ever more competitive spirit.

While the US and the USSR never managed to merge their lunar efforts, one example of international cooperation did emerge from those years: the Outer Space Treaty (OST) of 1967. This agreement provided some legal structure to the utilization of space, specifically barring the use of weapons of mass destruction and other strictly military activities. It also identified space and the bodies beyond Earth as a realm for peaceful programs and forbade territorial claims, much as the earlier Antarctic Treaty did for the South Pole. Since 1967, 105 nations have joined the OST fully, and twenty-six more have signed and are in the process of ratification. In the beginning, the primary concern was to find common ground for the two powers then entering space—the US and USSR—but since then, most other nations with spaceflight ambitions have joined.[92]

It was not until 1975 that international cooperation truly reached space with the Apollo-Soyuz Test Project (ASTP). After years of tricky negotiations and technical collaboration on the ground, the final flight of the Apollo program was launched to dock with a Soviet Soyuz in orbit. But Apollo-Soyuz (or as the Soviets called it, Союз-Аполло—with the Soviet spacecraft's moniker understandably coming first) was not so much a program as a political gesture. The single flight was an attempt to put the new policy of détente between the two nations into action in a neutral environment, or as Soviet leader Leonid Brezhnev put it, "The Soviet and American spacemen will go up into outer space for the first major joint scientific experiment in the history of mankind. They know that from outer space our planet looks even more beautiful. It is big enough for us to live peacefully on it, but it is too small to be threatened by nuclear

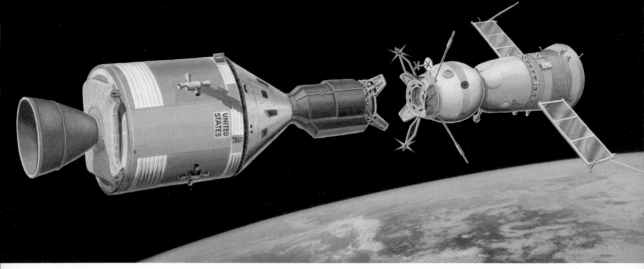

Artist's concept of an Apollo spacecraft docked with Soyuz. Image credit: NASA

war."[93] As a public relations gesture, Apollo-Soyuz succeeded, but no meaningful US-Soviet cooperation in space followed in its wake for almost twenty years.

In 1981, the American space shuttle began flying, and by 1984, NASA was planning its second space station, a modular outpost called Freedom. Freedom got off to a slow start, however, and would not take root as the International Space Station until the 1990s. Meanwhile, a new Soviet space station, called Mir ("Peace"), became operational in 1986. Each nation had a profound

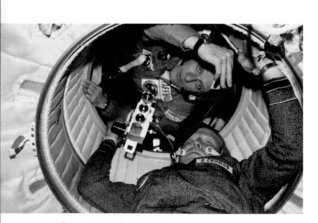

The first "handshake in space" between US and Soviet crews in docked Apollo and Soyuz spacecraft. At center, astronaut Tom Stafford and, at right, cosmonaut Alexei Leonov. Image credit: NASA

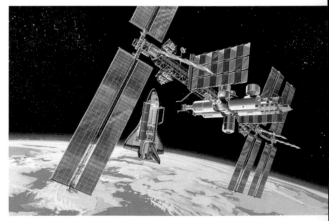

Artist's concept of the Reagan-era space station Freedom, with Japanese and European modules seen on the right. Image credit: NASA

presence in space—the US with its capable, heavy-lifting shuttle, and the USSR with its new modular space station—but it was clear that the two nations would benefit from cooperative ventures. Given the rising tensions between the two superpowers during the Reagan presidency, however, little was achieved.

It was not until 1994, three years after the fall of the Soviet Union, that the US and Russia once again flew cooperative crewed missions in space. A series of shuttle missions docked with the Russian space station between 1994 and 1998.[94] A total of eleven shuttle missions visited Mir, and a number of flight and research goals were achieved in this first true example of extended international cooperation in space.

In 1993, with a new president, Bill Clinton, in power, plans changed. NASA's space station would be called the International Space Station, or ISS. The US was working with a number of nations on the effort, as it had become clear that not only was the original program more expensive than the US wanted to support alone, but there would be positive public relations value to an international effort. Japan, Canada, and the eleven member countries of the European Space Agency served as the project's major partners.

Aside from the US, Russia was clearly the other heavyweight in human spaceflight and had both technological

The Russian/ex-Soviet Mir space station as seen from an American space shuttle closing to dock. A series of missions were undertaken with the shuttle visiting Russian crews on Mir. Image credit: NASA

expertise and flight hardware to contribute to a venture like the ISS. They had been planning a Mir 2 station for years, but had completed only the first module when that program was curtailed. With the downfall of the USSR in 1991 effectively ending the Mir 2 program, and the US hoping to add some stability to its relationship with the emergent Russian Federation, a bold cooperative initiative was suggested. The now-surplus Mir 2 core module was repurposed as the main power and propulsion unit for the ISS. A number of other Russian-built modules eventually joined US, Japanese, and European hardware to form the remainder of the station.

The completed International Space Station. Image credit: ESA

The internationalization of the space station did save the US some money over going it alone, though the actual price tag for the orbiting complex is surprisingly difficult to determine. The best estimate of the US contribution is roughly $78 billion, as of 2015, with the remainder of the station's costs, about $20 to 30 billion, coming from foreign governments.[95] In any case, the result is broadly perceived as a net win. NASA was spared the expense of developing a number of modules for the ISS, at the very least, and ongoing operational costs—such as ground control and astronaut training—are shared by the US, Europe, and Russia. This international effort has also generated much goodwill.

After the success of the ISS, many pundits knowledgeable about space, including *Apollo 11* moonwalker Buzz Aldrin and Mars exploration advocate Robert Zubrin, noted that internationalization is the only reasonable road ahead for human spaceflight. They cite the ISS as an example of how such cooperation can be effective—not just in cost sharing, but leveraging the expertise of many technologically advanced nations. This could be especially beneficial to a human Mars program.

As Bill Gerstenmaier, Associate Administrator for Human Exploration and Operations for NASA and the man in charge of human spaceflight, put it: "Mars is an unbelievable challenge we have in front of us. We are not going to do this as a single nation, or a single government; we will not do this as a single individual or a single corporation. We're going to take the absolute best from around the world, from the corporate and private sector, put all that together, and then we can accomplish this goal."[96]

NASA's official policy is to support international efforts in space, but it must do so while dealing with strict US legal constraints, known as ITAR—the International Traffic in Arms Regulation laws. Many US agencies are involved in ITAR, but in short, the law restricts international trade in anything that might pose a risk to national security. ITAR does not apply only to nations perceived to pose an immediate risk to this security, but to *any* foreign entity.[97]

ITAR rules are thorny. They loom rather ominously over every kind of technological cooperation between the US and other nations. Here's a description of ITAR and the chain of command for implementing it from the Electronic Code of Federal Regulations, Title 22, Chapter I, Subchapter M, Part 120:

§120.1 General authorities, receipt of licenses, and ineligibility.

Section 38 of the Arms Export Control Act (22 U.S.C. 2778), as amended, authorizes the President to control the export and import of defense articles and defense services. The statutory authority of the President to promulgate regulations with respect to exports of defense articles and defense services is delegated to the Secretary of State by Executive Order 13637. This subchapter implements that authority, as well as other relevant authorities in the Arms Export Control Act (22 U.S.C. 2751 et seq.). By virtue of delegations of authority by the Secretary of State, these regulations are primarily administered by the Deputy Assistant Secretary of State for Defense Trade Controls, Bureau of Political-Military Affairs.[98]

In practical terms, ITAR limits transactions involving obvious sorts of defense articles, such as weapons and ammunition, but can also apply to computer software and hardware, along with various commonplace technologies for which military use might be found. That includes many of the crucial components for a space program. To complicate matters further, different potential partners are covered by different restrictions. While it's not impossible for NASA to work with most international partners, it can be a very convoluted process.[99]

I asked Charles Elachi, the former director of NASA's Jet Propulsion Laboratory, for his thoughts on internationalism and space projects. JPL has worked with foreign partners for decades—indeed, the first cooperation in space science occurred during its robotic exploration programs of the 1960s—and Elachi has amassed vast experience in the complicated processes involved.

"I think international cooperation is essential. Most missions at JPL have some level of international collaboration, but it has to be done thoughtfully. I can give you two perfect examples. One is the Cassini–Huygens mission—the Europeans built the Huygens probe, and we did the rest. Another was the TOPEX Poseidon ocean observing system. We got twice the value for our money with that mission. So I think international collaboration is the way of the future."[100] In both projects Elachi cited, European partners tackled difficult technological challenges

The Cassini mission to Saturn was an international effort between NASA and the European Space Agency. NASA supplied the orbiting spacecraft, and ESA provided a probe to land on Saturn's moon Titan. The probe is the cone-shaped structure seen departing the bottom of the spacecraft. Image credit: NASA

and provided support during the mission, saving NASA a worthy percentage of the project cost—about $500 million on Cassini alone.[101]

Elachi limited his comments to unmanned missions, which, while technologically sophisticated, are not as complex as human spaceflight projects.

Another perspective, this one pertaining specifically to human spaceflight, comes from Aldrin, who has been quite vocal on the subject of international cooperation, particularly efforts between the US and China. In a 2015 talk at Yonsei University in South Korea, he said: "America should be able to help China, not compete with [them, but to] join the ambitious goal for Mars . . . When people work together, you can sometimes accomplish the impossible. Apollo could never have happened without the cooperation and efforts of many people working together for a shared goal." Aldrin added that the US should also be working with additional countries, including Korea and India.

The issues that ITAR attempts to address are very real, however. Russia and China, in particular, are known to be aggressively pursuing the militarization of space. The programs they have undertaken, or are suspected of pursuing, include the jamming or disabling of rival nations' satellites from the ground and the

Concerns about the militarization of space include anti-satellite (ASAT) weapons. An artist's concept of one possible system, considered in 1983, is seen here. Image Credit: US Department of Defense

interception and destruction of them in space.[102] In 2018, President Trump authorized the formation of a US Space Force, a separate branch of the military devoted to defending American assets in space, ultimately both military and civilian. Space Force is generally regarded as a necessary defense strategy against perceived foreign threats, but there are also concerns about a resultant escalation in space weapon capabilities.

Understandably, this worries some prominent thinkers and policy makers, including Scott Pace, executive secretary of the National Space Council.

"There are rising challenges from Russia and China and their counter-space capabilities," Pace told me in June 2017. While their military abilities currently exist primarily as anti-satellite weapons and the tools of electronic warfare, Pace sees these perils altering the motivation for human spaceflight as well. "Foreign threats are a more sustainable rationale for human spaceflight, and now that needs to be tied in to our geopolitical interest as well as our economic interest. So making sure that human space flight is aligned with longer-term US geopolitical goals, such as dealing with rising Russian and Chinese threats, is a big concern."[103]

Pace referred to the space race for context. "The Apollo model was about swaying the hearts and minds of developing countries that were coming out from decolonization, and was a contest between communist models and Western models. Which system would the developing countries align with? The space race was, to use a phrase from that time, 'part of the fluid front of the Cold War.' At the time, international leadership was about 'look at what we can do by ourselves, and see how capable and great we are.'"

Pace does see potential opportunities for international cooperation moving ahead, however. "Leadership today is about what can you get other people to do with you, not about what you do by yourself. And in order to get people to do things with you, these activities have to be

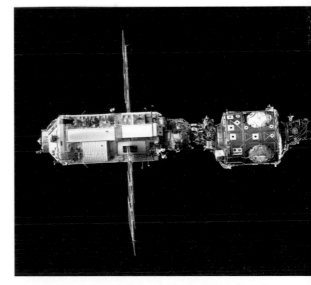

The first two modules of the International Space Station in 1998. To the left, the Russian Zarya module, and to the right, the US Unity module. Image credit: NASA

challenging and impressive, but in a way in which they can participate."

In Pace's estimation, Mars is too challenging a goal for the near future, and international partners may not have a lot to offer on the technological front yet. "But an international, US-led return to the moon is one goal that provides more international partnership opportunities while still contributing toward Mars as a much longer-term goal. If we decide to do an international return to the moon and include others, they will come with us," he says.

Given the restrictions caused by ITAR, overall security concerns, and the practical limitations of international partnerships, it's important to work out a constructive framework for meaningful cooperation in space. We've already had a good run with the ISS, so we know that international collaboration can work. We simply need to find the proper path forward, one in which all partners can share in the bountiful returns of technological development, education, privatization of spaceflight, and even national pride.

It's a lofty-sounding ideal, but not an impossible one. It will, however, take a lot of trust, negotiation, and discipline to accomplish.

As we have seen with the ISS, many nations other than the US and Russia have capabilities in space technology, and their strengths are notable. The accomplishments of these national programs and the challenges they face can provide us with insight on where the US might find cooperation the most beneficial.

EUROPE

Other than a brief collaboration in missile technology between the Soviet Union and China, which was purely military in nature, international cooperation between national space powers was begun in Europe in the 1960s. Britain's space efforts kicked off in 1952, and many of the major European nations laid the foundations for their own programs over the next two decades. The current French space agency, *Centre national d'études spatiales* (CNES), was founded in 1961, followed by the German Aerospace Center, or *Deutsches Zentrum für Luft- und Raumfahrt e.V.* (DLR), in 1969. In 1975, the spacefaring powers in Europe combined to found the European Space Agency (ESA).

By 2008, ESA member states had flown cargo to the International Space station with their uncrewed Automated Transfer Vehicle (ATV), and contributed a science module to the ISS called *Columbus*,

Artist's concept of the European Space Agency's HTV cargo spacecraft approaching the International Space Station. Image credit: ESA/D. Ducros

along with the windowed cupola installation. ESA also has a robust planetary exploration program and has been notably successful with their probes to Mars, comets, and outer solar system targets.

In 2016, the director-general of ESA, Johann-Dietrich Wörner, announced the ambitious goal of creating an international "Moon Village" in the 2020s. ESA plans to accomplish this in cooperation with private industry and the national space programs of the United States, Russia, China, and possibly other nations. ESA—and in particular its largest national components, France and Germany—is preparing for an increased role in both robotic and human spaceflight. I spoke with Pascale Ehrenfreund,

chair of the DLR executive board, to gain some insight into ESA's projects and goals for the next two decades.[104]

"With respect to human spaceflight for the next ten to twenty years, the ISS remains the primary destination. Germany is investing $900 million in order to secure the German component, which represents a large part of the European investment in the space station. We see this as an important responsibility, and this includes doing a lot of science on ISS, at least through 2024. Looking ahead the next ten to twenty years, we must consider what comes after the space station. We're coming up to a crossroads because we don't know exactly how space exploration will be moving ahead in the United States—this is an international effort.

"We are deciding what roles we can play moving ahead in the international program . . . this is something which we really want to concentrate on. Do we stay in low Earth orbit after the space station with smaller stations? Or do we go back to the moon?"

I asked Ehrenfreund for her thoughts regarding the European private sector—along with China, Europe would seem to be best suited as the source for the next SpaceX or Blue Origin. Her remarks were enlightening.

"We are not so fortunate to have billionaires willing to invest in space as you see in the US. The conditions are very different in Europe. We have very little venture capital—we tend to be risk averse and don't have big chunks of public money to invest in the private sector the way NASA has by funding SpaceX."

NASA has spent billions in its new cooperative ventures with SpaceX, Northrop Grumman Innovation Systems, and others. Europe has not yet followed this course with simlar investment in its private sector.

"It is clear, however, that all the nations of Europe are watching what is happening on the other side of the Atlantic and thinking about how we could also activate a market for space applications," Ehrenfreund continues. "Ultimately, we will have to find our own answer to the Space 2.0 question."

JAPAN

Japan's earliest organized space efforts began at the University of Tokyo in the 1950s. Japan's spaceflight programs were consolidated in 1969 with the formation of the National Space Development Agency (NASDA). In 2003, NASDA merged with two other Japanese space entities to become the Japanese Aerospace Exploration Agency, or JAXA.

Though Japan has studied and considered many approaches to human spaceflight via capsules and even a spaceplane, the Japanese government was ultimately unwilling to invest the heavy funding required to support those efforts. They instead financed a series of unmanned launch vehicles. With this capability, the Japanese have been quite active in satellite launches and planetary exploration. Other parts of its program remain active, as well;

Japan's astronauts have flown aboard the US shuttle and to the ISS.

Hiro Iwamoto, the director of JAXA's Washington, DC, office, sees a bright future for Japan's space program.[105]

Japan's International Space Station HTV cargo vessel prepares to launch from Japan. Image credit: ISAS/JAXA

"JAXA has [many] goals for the next ten to twenty years. We have been developing our H3 launch vehicle. The first launch of H3 is scheduled in 2020, along with improving our Epsilon rocket. Our goal is to make these rockets [operational] and to also be competitive in the commercial market." The H3 is a medium-sized rocket, similar in capability to the American Atlas.

JAXA has also launched cargo to the ISS with an uncrewed supply spacecraft called the HTV and is now working on a successor called HTV-X to press farther into space. "Besides providing logistical support to the ISS, the HTV-X could be used for human interplanetary transportation," Iwamoto says.

JAXA is also active in robotic space exploration. In 2005, its Hayabusa probe rendezvoused with an asteroid, returning a small sample to Earth in 2010. Hayabusa-2 was launched in 2014 and will return new asteroid samples in 2020. The SELENE mission orbited the moon in 2007, capturing and transmitting the first high-definition images and creating new, highly detailed maps of the lunar surface. Japan is also planning a Mars mission that will explore the planet's moons, Phobos and Deimos, and continues to monitor IKAROS, a solar sail spacecraft launched in 2010. The solar sail craft, which uses the force exerted by photons from sunlight for propulsion, is a design

Artist's concept of Japan's Hayabusa asteroid rendezvous mission in 2005. Image credit: JAXA

long researched by the US and others, but pioneered in space by Japan. It may prove to be the most efficient way to move non-human payloads across the vast distances between planets.

Besides JAXA's cooperative efforts with NASA on the ISS, it has also participated in many robotic exploration missions with NASA. JAXA and NASA are currently working together on a Mars program called the Martian Moons Exploration mission that will launch in 2024 and return the first sample of the Martian moon Phobos to Earth. They are also collaborating on the X-Ray Astronomy Recovery Mission (XARM), an orbiting observatory currently scheduled for launch in 2021.

INDIA

India's national space agency is called the Indian Space Research Organization, or ISRO. It was established in 1969 and launched its first satellite in 1975 aboard a Soviet booster. India began launching satellites on domestically made rockets in 1980 and since then has increased the pace and type of operations, including the launch of foreign commercial satellites.

Like the ESA, India's ISRO expanded operations as the space race was winding down, with the stated goal of peaceful exploration. And, as is the case with China, the Indian national space program was seen as a way to promote education and technological excellence in a rapidly developing country. The level of engineering and scientific expertise required for space exploration has proved to be a strong incentive for schools and their students to develop such skills.

India, like all countries that commenced spaceflight after the 1960s, benefited from the technical innovations forged by the US and USSR during the space race, much as both those powers benefited from the developments in rocketry by Germany in World War II. But that is not to take anything away from India's achievements—it has evolved its spaceflight technology with domestic talent and has the added benefit of far lower labor costs, allowing the nation to achieve impressive feats in space for less than the Western space powers. This puts the country in a prime position for international cooperation in the future, providing hardware and expertise for spaceflight at lower cost.

A model of India's PSLV rocket on display at the Science City, Kolkata. Credit: Wikimedia Commons/Biswarup Ganguly

Artist's concept of India's Mangalyaan Mars orbiter. Image credit: Wikimedia Commons/ Nesnad

Among India's numerous contributions to science—some on its own rockets, and some working with other agencies such as NASA—the country was the first developing space power to successfully reach Mars. India scored a major coup when its Mars Orbiter Mission (MOM, also known as Mangalyaan) successfully reached orbit around the Red Planet in 2014. While the probe is primarily an engineering mission testing robotic spaceflight technologies, MOM is returning valuable scientific data. Due to a very wide orbit, the probe is collecting and transmitting full planetary views long coveted by scientists. Such imagery has had a profound impact on the study of Martian weather cycles and geology.

To put this achievement into context, other than the United States, only Europe and India have succeeded in sending spacecraft to Mars. The Soviet Union tried to reach Mars dozens of times beginning in the 1960s, with Russia picking up the challenge since the founding of the Russian Federation in 1991, with only partial success in each case. Japan, Britain, and China have also attempted to send hardware to Mars, usually in cooperation with Russia, but have failed every time. As they say at NASA's JPL, Mars is hard.

India's robust list of future mission plans includes a continuing series of Earth observation satellites; a lunar orbiter called Chandrayaan-2 (Chandrayaan-1 operated in lunar orbit for almost a year starting in 2008); a solar research orbiter called Aditya, scheduled to launch in 2019; a Venus orbiter in 2019; and a Jupiter probe that is yet to be scheduled. These are ambitious goals for a country still in the early phases of space exploration and indicate both confidence and the willingness to fund their goals despite some domestic opposition. There are prominent voices in India that declaim such programs when there are so many problems facing the developing country, but the Indian government has chosen to move forward in space.

With regard to human spaceflight, there has been slow but ongoing movement toward Indian crewed flights into space, which would make them the fourth country to do so, after the US, Russia, and China. An unmanned test capsule was successfully flown in 2014 to investigate various systems and techniques needed for human spaceflight. The capsule was small, only about the same mass as the US Mercury capsule of the 1960s, but engineering data gained from this mission provided an important first step in scaling up to a larger, crewed spacecraft now planned for launch in 2022.

India has cooperated with a variety of international spaceflight partners, working closely with NASA on planetary probes, providing launch services for private companies in the US and Europe, and collaborating with the Soviet Union and later with Russia. India recently entered the commercial launch services market and has led the way in launching multiple smaller payloads in one flight—a February 2017 Indian launch carried a record 104 small satellites into orbit on a single rocket.

One of the most remarkable aspects of Indian space endeavors is its cost. The MOM Mars project is estimated to have cost about $73 million, making it by far the least expensive mission to that planet. The MOM program benefited from the reduced costs of labor and expertise in India. (China enjoys similar advantages but does not publicize its budgets in this detail.) India has accomplished much with its $1.5 billion annual space budget—which is about one-third to one-half of the budgets in China, Russia, Japan, and Europe, and pales next to America's $19 billion annual NASA budget.

Based on costs alone, Western nations may become more frequent clients of Indian expertise, in much the same way that Asia has taken over much of the manufacture of consumer goods in the past few decades. The upside of this trend is that Western partners can build and launch spacecraft for lower costs, while proactively engaging the international space sector. This would also serve to promote high-tech jobs and education in some of the partnered nations. There are issues of "offshoring" to be considered, but it may be an inevitable part of international cooperation in space.

Like the US and Europe, India is interested in developing commercial spaceflight companies domestically. So far this has been limited to producers of subsystems, such as satellites and flight-ready electronic components. No private rocket companies along the lines of SpaceX or Blue Origin have yet arisen in India (such efforts are just beginning in China). Creating a thriving private spaceflight sector, a natural goal for an aggressively emerging technological society like India, will require a lot of flexibility within both the ISRO and the Indian government. Substantial private investment will be key. But given the growth of India's tech sector, and the Indian economy in general, it would seem to be an inevitable expansion, one similar to what the country has accomplished in other technical fields.

CANADA

The Canadian Space Agency was created fairly recently, in 1990. Its roots can be traced back to a number of small rockets developed by the Canadian government in the preceding decades, but these were suborbital and military in nature. A Canadian satellite program also existed, launching on American rockets. The CSA's budget is small—only about a half-billion dollars annually—so rather than attempting to create an independent launch capability, the country fosters strong partnerships with ESA, NASA, and other national space agencies. One of its most visible contributions was the Canadarm, the robotic manipulator arm flown on the space shuttle and the ISS. NASA approached Canada in 1975 about partnering on this component, and the Canadarm was a standard feature on the shuttle across its operational lifetime. A Canadarm 2 was installed on the ISS in 2001 and offers greater capability. It is longer, able to handle more mass, and has greater flexibility via more arm joints.

A number of Canadian citizens have also flown on the shuttle and ISS over the years, the most famous of them being

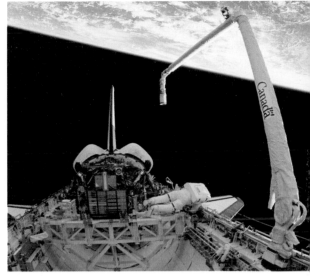

The Canadarm is seen on the right, rising from the space shuttle's cargo bay. Image credit: NASA

Chris Hadfield, who entertained millions on Earth by playing his guitar and singing from the space station in 2013. Various NASA missions and a number of ESA's planetary missions have included Canadian scientific instrumentation. Canadian companies have also invested in US space companies, with US/Canadian company Maxar Technologies Ltd. acquiring Space Systems/Loral (now SSL), a leading US satellite manufacturer, in 2012.

OTHER CONTENDERS

There are a number of other small national space agencies, with new ones popping up every few years. Iran, North and South Korea, Saudi Arabia, and even tiny Bangladesh and Luxembourg (with a population of just 535,000) have announced their intentions in spaceflight. Some are building rockets, and some are building satellites and other space hardware.

Most notable may be Luxembourg and the United Arab Emirates. Luxembourg operates the Luxembourg Space Cluster, an association of high-tech companies active in telecommunications and other space-related specialties. The country announced its intention to invest in space mining companies, most famously via its association with the US-based companies Deep Space Industries and Planetary Resources.

The UAE has taken quite a different tack. The United Arab Emirates Space Agency (UAESA) was formed in 2014, with the goal of creating a world-class space sector. The agency built an Earth observation center in Abu Dhabi and has launched a number of satellites on other countries' rockets. The UAESA has since announced an ambitious plan to launch a Mars orbiter in 2020 in collaboration with a number of US universities and other international partners. The spacecraft will feature a suite of instrumentation designed in the UAE to study the Martian atmosphere. The Emirates' space efforts will be a profound driver of domestic technology and education.

More remarkably, in early 2017, the UAE declared its intentions to build a city on Mars by 2117. It plans to work with international partners to encourage human flights to Mars over the next few decades, while gradually building up infrastructure on the Red Planet. Anticipated benefits beyond national pride are the strengthening of the Arab tech sector and greater engagement of young engineers, scientists, and technicians. UAESA's overarching goal is to spearhead efforts to develop food, energy, and other logistical resources on Mars, leading up to the construction of the city on the planet's surface. It's an ambitious plan—a century of continuous technological development—that relies upon the relative stability of leadership and continuity of goals provided by a monarchical form of government.

OVERSEAS SPACE ENTREPRENEURS

We are just beginning to see the emergence of space-focused entrepreneurs in the mold of Musk and Bezos internationally. There have always been individuals around the globe investing in aerospace, but to date no international billionaires have stepped up to create a foreign SpaceX or Blue Origin. Plenty of private capital exists in places like China, India, and Russia. With some changes in regulation, proper government incentives, and the right potential collaborations with US entities, wealthy individuals in other countries may move toward more robust public-private sector partnerships, eventually spawning private entities doing big business in the space sector. With lower labor costs, highly trained and increasingly educated workforces, and lower levels of government regulation, rapidly developing economies, such as India, could soon give the rest of the world a run for their money in spaceflight. Private launch companies are just getting started in China (see Chapter 13).

Economic gain and scientific curiosity are not the only motivations for countries such as India and China, of course. Although their paths are different, with China moving more aggressively toward human spaceflight and India currently focused on robotic exploration, their overall progress resembles that of the US and USSR in the 1960s, with strongly nationalistic overtones to their development decisions. The United States, Europe, and Russia have been flying into space for decades, but are still driven to limit cooperation and keep their programs largely domestic in nature for reasons of both national pride and national security. India and China, both fairly new to spaceflight, have articulated their nationalistic goals clearly.[106]

Despite this, the best outcome for these countries' space programs will be reached through true international cooperation, with each party contributing in accordance with its individual strengths toward the completion of a collaborative vision. How this can be achieved, and in what time frame, is a work in progress; despite the successes of the ISS, it may be too soon for true globalism in space. But as these international programs advance, and their goals converge, efficiency will demand it.

Image credit: James Vaughan

CHAPTER 12

THE RUSSIAN JUGGERNAUT

And then there is Russia. Of all the nations that have aimed for the heavens, Russia has traveled along the most daring yet convoluted path. The Soviet Union was first into space with Sputnik in 1957, the first to orbit a human in 1961, and the first to launch a space station in 1971. Its Soyuz spacecraft, initially designed to beat the Apollo flights to the moon, has become the longest-lived and most successful space taxi in history. The USSR's early planetary exploration programs were sensational throughout the 1960s, with ten Venera probes reaching Venus over a twenty-three-year period. The only planetary target that eluded them, and continues to, has been Mars.

Let's take a moment to inventory some of the Soviet Union's (and Russia's) substantial achievements:

- First Earth satellite—Sputnik, 1957 (robotic)
- First animal in space—Sputnik 2, 1957 (a dog in orbit)
- First lunar impact mission—Luna 2, 1959 (robotic)

Image from the Soviet Venera 13 on the surface of Venus in 1982. Image credit: NASA

- First images of lunar farside—Luna 3, 1959 (robotic)
- First Venus flyby—Venera 1, 1961 (robotic)
- First human in space and Earth orbit—Vostok 1, 1961(human)
- First Mars flyby—Mars 1, 1963 (robotic; spacecraft failed before flyby but managed to pass the planet)
- First woman in space—Vostok 6, 1963 (human)
- First spacewalk—Voskhod 2, 1965 (human)
- First completely automated rendezvous and docking in space, 1967 (robotic)
- First circumlunar flight—Zond 5, 1968 (robotic, but carried reptiles and insects)
- First successful landing on Venus—Venera 7, 1970 (robotic)
- First robotic rover on the moon, Lunokhod 1, in 1970
- First crewed orbital space station—Salyut 1, 1971 (human)
- First soft landing on Mars—Mars 3, 1971 (robotic; operated for less than a minute after landing)
- First cargo supply missions to a space station, 1978 (robotic)
- First long-term, modular space station—Mir, 1986 (human, operated for a decade)
- First fully automated flight of a space shuttle—Buran, 1988 (robotic, but human-rated)
- Development of multiple large launch systems, a number of which continue to operate today, delivering both Russian and international payloads[107] (human and robotic)

Russia has had its share of disappointments as well. While its early human spaceflight accomplishments were impressive, its massive attempt to beat the Americans to the moon—either by sending cosmonauts to loop the moon before the US Apollo 8 orbited that body in 1968, or landing a lone cosmonaut there before Apollo 11 arrived in 1969—was plagued by a series of disastrous launch failures.

The successes outlined above occurred largely in the Soviet era, which ended before the Mir space station—launched in 1986 while the USSR still existed—ceased operations in 1996. As the country transitioned to the Russian Federation, many space program plans from the Soviet era were carried forward and new ones added, while others were abandoned outright.

Today, Russian space efforts are cash-strapped and plagued by intergovernmental rivalries. Russia has found it challenging to maintain a presence on the ISS and to continue its robotic exploration programs. The rise of private launch companies in the US, especially those such as SpaceX and Boeing that can carry humans to space, further challenge Russian efforts. In the future, each astronaut who flies to the ISS aboard a SpaceX Dragon 2 capsule or Boeing's Starliner, instead of the Soyuz, represents a roughly $82 million reduction

The Soviet Union's massive N1 moon rocket was launched four times and failed four times. Image credit: NASA

A Russian Soyuz rocket launches to the International Space Station with a Soyuz spacecraft containing Russian and US crewmembers in 2012. Image credit: NASA

in foreign revenue for Russia.[108] The loss of this revenue will deprive the Russian space agency of a substantial portion of its yearly budget—as much as an eighth or more.

One place Russia has continued to shine is with its robust launch vehicles, primarily the Soyuz and Proton boosters. Both have reached a point of high reliability and both launch frequently, carrying heavy payloads into space.[109] And the Soyuz spacecraft, first flown in 1967, has matured into the world's safest and most frequently used human space transporter. Russia continues to develop a replacement for the aging Soyuz, a new spacecraft called Federation. This larger capsule is much more similar in design to the US Apollo and Orion capsules than it is to the Soyuz. Federation will fly humans just as the Soyuz does, but it's also planned to have a fully automated variant for cargo delivery to the ISS and a possible future Russian space station. Roscosmos, the state corporation overseeing the Russian space program, has been investigating potential replacements for the Soyuz booster as well. One, the Angara, is a modular rocket that has been under development since 1992. The Angara will be able to carry payloads of various masses depending on the rocket's configuration—its modular design allows it to be customized to carry lighter or heavier payloads. There are also discussions of a super heavy-lift rocket,

currently unnamed, that could enter service in the early 2020s. It would launch larger, heavy spacecraft, possibly including crewed lunar flights.[110]

Russia, along with other launch providers such as China and India, faces a major challenge as it moves ahead with these plans: reusability. With all the major commercial launch providers in the US developing reusable launchers and capsules, one-use boosters, unless created far more cheaply than they are today, will become a liability. In this area, SpaceX and Blue Origin have set the new standard, and other launch providers will have to follow suit to remain competitive.

In 2016, Vladimir Putin brought most of the Russian space industry back under state control, similar to the way it operated in the Soviet era. This move was intended to increase efficiency and reduce corruption. Money, however, remains a problem. While NASA struggles to finance new programs with a flat budget of between $18 billion and $19.5 billion per year, in 2015, the Russian government approved a $20 billion budget—to fund the next *decade* of space activity. That's about $2 billion per year, or one-tenth of what NASA spends. While costs in Russia are lower than in the US—technicians and engineers generally earn between $200 and $300 per month—the differential does not close the gap. As John Logsdon, former director of the Space Policy Institute at George Washington University, put it, "Their budget is not adequate to maintain a world-class space effort across the board."[111]

The situation was put in a starker light by space historian Asif Siddiqui: "Roscosmos has been beset with corruption, mismanagement, and crony capitalism that is the hallmark of the larger post-Soviet economy. In a tech sector that needs to meet very high standards, these problems have led the workforce on the ground to cut corners."[112]

In November 2017, a Soyuz upper stage failed at liftoff, after an earlier launch failure in 2016 and *three* in 2015. These do not necessarily indicate flaws in the specific hardware used to shuttle crews to the ISS, but they surely indicate larger quality-control problems at Roscosmos.[113]

Since the US has relied on Russia since 2011 for crew access to and from the ISS, NASA watches Russian launch activities carefully. With SpaceX and Boeing running late in their programs that would replace these Soyuz flights, it's important to confirm that the Russian rockets are reliable. Nobody in either country wants to lose a crew.

So, what lies ahead for the Russian space program? Future challenges include funding from the Russian government, the country's relationship with the US (which has been particularly problematic in recent years), a continued lack of investment in the tech sector and higher education, and a decline in population that has hit high-tech fields harder than most. In contrast to the Soviet era, when the USSR spent whatever it took in its attempt to beat the US to the moon, today the European Union, China, and even Japan spend more on their space programs annually than Russia. Russia is relying on older designs and hardware to struggle forward in spaceflight—not a promising prescription for future success.

Given the Russian program's financial limitations, its success may depend upon increased international collaboration. ESA is the most likely partner, and India and other national programs may choose to buy or license Russian technology, as the Chinese did when they licensed the Soyuz spacecraft design to shortcut development time. Depending on the political climate, the US may or may not find it comfortable to partner with Russia once ISS operations conclude. The US currently plans to cooperate with Russia on the planned Lunar Orbiting Platform-Gateway, or LOP-G (see Chapter 16).

There has been US private-sector involvement in the Russian space industry over the years as well. As noted in Chapter 9, ULA has used Russian engines for its Atlas V rockets since 2002—but will soon be switching over the rocket engines build by Blue Origin. Congress passed an amendment in 2016 restricting the use of Russian rocket engines on US national security launches, and the bulk of Atlas V launches are sold to the military. Such private sector collaborations are all but certain to grow more complicated in the years to come.

While Russia continues to fly aging hardware that it struggles to replace, China has emerged as the most powerful and aggressive new space power. Though the country has been working on missiles and larger launch systems for decades, it has only proven itself in the past two decades to be a major player. Progress has been swift, and a small private spaceflight sector is emerging. Let's take a closer look at the newest global power in human spaceflight: China.

A Russian Soyuz spacecraft docked with the International Space Station. Image credit: NASA

Image credit: James Vaughan

CHAPTER 13

THE CHINA WILDCARD

Early in the twenty-first century, China joined the US and Russia in human spaceflight. Since 2003, China has flown six crewed missions to orbit and lofted two small space stations. A third, larger station, modular in design like Mir and the ISS, is expected to debut sometime between 2019 and 2022, and a human landing on the moon is in China's sights. Like the US and Russia before it, China has tested and improved its hardware incrementally, with increasingly ambitious goals for each flight.

The Chinese space program started back in the 1950s, at about the time the Soviet Union was preparing to launch Sputnik. It originated from the same source as both the Soviet and US space programs—a military need to develop nuclear missiles. Early on, the Soviet Union shared technology with the Chinese, and progress was swift. Then, China and the USSR parted political company in 1960. The Chinese government shelved any plans involving human spaceflight, but only for a short time.

In the mid-1960s, Chinese leadership decided that it was time for their country to once again consider sending people into space. But then, in 1966, the Cultural Revolution ignited, and soon it was ripping through the nation. Mao Zedong, chairman of the Communist Party, instructed Chinese youth to purge the society of "impure elements." The associated upheaval of both business and industry stalled national technological development. Though the Cultural Revolution continued through 1976, the worst of the disruption was over by 1969, after Red Army troops were dispatched to quell the excesses of the Red Guard, the student groups Mao had inflamed just three years earlier. With both the military and industry less fearful of disorder, China's first orbital satellite was finally able to launch in 1970.[114]

Subsequent progress was slow, but more launches followed in the next few years. In 1971, the audacious goal of sending Chinese astronauts (commonly referred to as "taikonauts" in the West) was set for 1973, just two years hence. The spacecraft was to be called Shuguang ("Dawn"), and was similar in general design to NASA's Gemini spacecraft. But internal political struggles killed the program before it could start in earnest.[115]

Upon Mao's death in 1976, and the end of the Cultural Revolution, human spaceflight resurfaced as a possible priority for China, but it ranked far below missile development in importance. Ideas

for a human spaceflight program were not revisited seriously until the mid-1980s, after which they evolved into today's Shenzhou program.

The main headquarters for the Chinese space agency, the China National Space Administration (CNSA), is located near Beijing, with the Shenzhou project headquartered within the new Space City complex just outside Shanghai. This expansive facility operates in conjunction with more than a dozen major space and military facilities scattered across the country. The effort is reminiscent of the Cold War military-industrial partnerships of the United States and the Soviet Union, in which massive programs were assembled from components built by hundreds of thousands of people in geographically diverse locations, sited to capitalize on their intrinsic strengths—for example, being sufficiently remote to allow for the live-fire testing of rocket engines—and in other cases to meet political and economic requirements.

Though represented as a civilian program, Chinese space efforts operate under the influence of the Chinese military, in contrast to America's civilian space agency. Because of this, it is less transparent and accessible than NASA, but the results that are announced demonstrate an impressive agenda.

China has developed a series of rocket boosters called Long March since

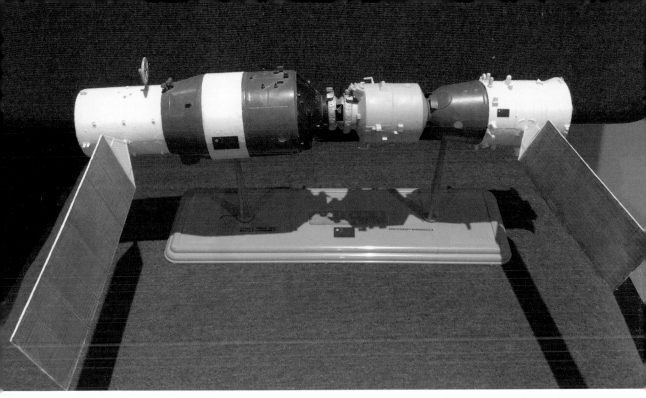

A model of the Tiangong-1 station, to the left, docked with Shenzhou, to the right. Image credit: Wikimedia Commons/Leebrandoncremer

the 1960s, and new iterations continue to carry the name forward. Long March refers to the Communist Army's retreat in 1934–35 from Republican forces in the Chinese civil war and has long been a rallying-cry of the People's Liberation Army (PLA). Long March rockets have carried all crewed Chinese missons and quasi-civilian space hardware into orbit.

In 2003, the flight of Shenzhou 5 was the first to carry a taikonaut after four uncrewed test flights. Since then, the Shenzhou program has progressed through a number of increasingly aggressive phases, just as the US and Soviet programs of the 1960s did. Test flights

in orbit, then a spacewalk (EVA) by a taikonaut, followed by the launching of China's first small space station, Tiangong-1, in 2011. Hardly larger than the Shenzhou spacecraft itself, the habitat was visited by three crews through 2013. Chinese ground controllers subsequently lost contact with the station. It reentered the Earth's atmosphere harmlessly in 2018.

Tiangong-2 was launched in 2016. A larger and more capable station, it has successfully hosted two taikonauts, who arrived in Shenzhou 11 and stayed for a month, as well as a robotic resupply craft that successfully docked and refueled the station twice. The first module of a larger, modular space station will be launched in 2020 or soon thereafter, to be joined by other modules once in orbit.

China has developed large and capable rockets, which now carry payloads for international customers on a regular basis. China is second only to the US in terms of flight frequency. With a few commercial launch enterprises beginning to gear up, China may soon surpass the US in frequency, but it remains to be seen how quickly Chinese commercial space enterprises can accomplish their goals.

In late 2017, a Chinese company called Link Space announced its intention to build a small, reusable rocket to service the commercial launch sector outside of the Chinese space agency. The vehicle is called New Line 1 and is a single-stage rocket that would be capable of carrying only a fraction of what a Falcon 9 can—about five hundred pounds. More importantly, Link Space represents an experiment in the private sector in China, where spaceflight is a domain carefully controlled by the government and the military. If Link Space proves successful, it could open up the floodgates of Chinese entrepreneurial activity. What involvement the Chinese military might have in Link Space remains unclear, but the company has publicly positioned itself as a private start-up. The cost to launch a newly manufactured New Line rocket

would be less than $5 million, and probably much less once it has been recycled. The company has flown more than 200 tests of smaller rockets that were capable of hovering and landing, in the same way that SpaceX gained experience with its own experimental flights before offering the Falcon 9 for commercial launches. Link Space expects to begin commercial service in 2020. This may be just the tip of the entrepreneurial iceberg; other Chinese start-ups are eyeing space launch and satellite markets, and are doubtless watching Link Space carefully to gauge its progress and success. At least one smallsat launcher outfit, called One Space, has been jump-started with official support from the Chinese defense establishment. This could be a foreshadowing of public-private arrangements in China similar to NASA's COTS program.[116]

Possible entrepreneurial activities aside, when viewed through one lens, the Chinese space program can be seen as a cloning of the space race activities of the US and USSR. Seen through another lens, however, the modern Chinese space program can be viewed as an intelligent, well-planned progression through the phases of spaceflight testing that proved successful for those who went before, with an early embrace of at least partially private rocket companies. While clearly nationalistic in origins—just as the programs of the space race were—China

China's Shenzhou spacecraft. Image credit: Rod Pyle

has intelligently improved preexisting, proven technology with Shenzhou, having purchased a number of Soyuz spacecraft from Russia after the fall of the Soviet Union. They have substantially modified the Soyuz; the Shenzhou is longer, heavier, and somewhat more capable, but the Chinese design builds on one that has enjoyed a fifty-year record of reliable and safe spaceflight. In contrast, the Tiangong space stations have been developed entirely in China—a sure sign of increasing self-reliance in space technology.

These achievements have not been without criticism from abroad. Despite the economic partnership and robust business exchanges between China and the United States, perceptions of

intellectual "borrowing" and outright theft of advanced technologies across a broad spectrum (via both traditional espionage and computer hacking) have crippled potential Sino-US partnerships in space. There had been talk in the 1990s by NASA and its international partners of including China in the multinational family supporting the International Space Station. George Abbey, the director of the Johnson Space Center from 1996 through 2001, and Leroy Chiao, a shuttle astronaut, were among the supporters of this idea. But the United States Congress blocked this initiative in 2011 by inserting prohibitive language in NASA's funding bill. Even without that roadblock, any meaningful future partnership between the US and China will have to proceed slowly just to navigate ITAR regulations.[117] The net effect is that much of what might be accomplished through US cooperation with China has been stymied and slowed. China has continued on its own, aided by Russian technology and support. China has also been in discussions about collaboration with a number of other national space programs, primarily the European Space Agency. Its program remains robust and fast moving.

On the robotic space exploration side, China has sent orbiting probes to the moon and a robotic rover to the lunar surface. The Chinese host an armada of Earth-orbiting satellites, are planning a lunar sample return mission, a lunar far-side landing (something not yet achieved by any country), a multi-landing robotic research complex near the lunar south pole, and plan a crewed landing there in the mid to late 2020s. They also intend to send an orbiter and rover to Mars in the early 2020s and have discussed the possibility of a human mission to that planet by 2040.

While the Chinese space program's budget is currently far smaller than the US's—about $2 billion compared to NASA's roughly $19 billion—its expenses are lower. Add to this the fact that the Chinese are not supporting many "legacy" missions; in addition to the ISS and the development of the SLS and Orion, NASA's funding must include an outlay for the forty-one-year-old Voyager program, which is still transmitting data, and the Mars Exploration Rover program, now in its fourteenth year on the Red Planet, and close to two dozen other active missions in Earth orbit and beyond. Neither are the Chinese supporting an expensive permanent space station—Tiangong-2 is currently mothballed. As of today, China's budget is targeted at new and emerging programs, such as the upcoming modular space station. Combine this with the nation's economic power, and the country has the potential to advance quickly in human spaceflight research.

Details of China's plans for the future come largely from official documents, the most recent of which was a long report released in 2016 that specified national goals for spaceflight and development.[118] The preamble to the report stated China's intentions:

> To explore the vast cosmos, develop the space industry, and build China into a space power is a dream we pursue unremittingly. In the next five years and beyond China will uphold the concepts of innovative, balanced, green, open, and shared development, and promote the comprehensive development of space science, space technology, and space applications, so as to contribute more to both serving national development and improving the well-being of mankind.

This seems to be a clear call for the peaceful development of space, though Western observers such as Scott Pace would appear to doubt its sincerity. The report stated that China supported "open development," combining independence and self-reliance with a willingness to join the outside world and participate internationally in scientific advancement. This would seem to indicate a continuing interest in collaboration with Western powers and possibly, one day, the US. Which specific Asian countries might be included in the circle of cooperation was not specified.

Next came a discussion of technological, scientific, and program-related plans. China's goals for the future are as wide-ranging as they are ambitious. Again, from the report:

"We will develop and launch medium-lift launch vehicles which are non-toxic and pollution-free, improve the new-generation launch vehicle family, and enhance their reliability." The report continues, "China will conduct research into the technologies for low-cost launch vehicles, new upper stage and the reusable space transportation system between the earth and low-earth orbit."

This, to a large degree, shadows what is being done by SpaceX and Blue Origin with regard to reusability—a smart move in the coming Space 2.0 economy. The report then lays out some objectives that look beyond the moon and a landing on Mars: "[The program] will conduct further studies and key technological research on the bringing back of samples from Mars, asteroid exploration, exploration of the Jupiter system, and planet fly-by exploration."

China's planetary exploration plans rival those of the US and Europe for the moon and Mars, though they are less clear with regard to the outer planets such as Jupiter and Saturn.

While announcing an interest in international cooperation, the report goes on to identify more nationalistic goals:

To build China into a space power in all respects, with the capabilities to make innovations independently, to make scientific discovery and research at the cutting edge, to promote strong and sustained economic and social development, to effectively and reliably guarantee national security, to exercise sound and efficient governance, and to carry out mutually beneficial international exchanges and cooperation; to have an advanced and open space science and technology industry, stable and reliable space infrastructure, pioneering and innovative professionals, and a rich and profound space spirit; *to provide strong support for the realization of the Chinese Dream of the renewal of the Chinese nation,* and make positive contributions to human civilization and progress [italics mine].

"Renewal of the Chinese nation" can represent a lot of things and, in this context, almost certainly includes a leading role in space exploration. The country's recent record of following through on announced programs and achieving stated goals over time is impressive. The Chinese political system has long allowed for presidents to serve for ten-year terms. This has recently been amended to allow presidential terms to extend for an indefinite period. These long terms, and the relative stability of the Chinese government, have resulted in the reliable execution of long-term space projects in the twenty-first century. With no real challenge from opposing parties, the Communist Party can plan what it wants to do and stick with that plan.

With China incrementally approaching rough parity with established space programs in terms of orbital spaceflight and planetary exploration, there is an increased interest in cooperation with the West. Looking at this through a historical lens, cooperation on robotic space missions has provided a soft entry into partnership in spaceflight before. This was the case between the US and the Soviet Union in the 1960s, and with Europe starting in the 1980s. I turned to the director of solar system exploration, Jakob van Zyl, at NASA's primary planetary exploration organization, the Jet Propulsion Laboratory, for comment.

"Working internationally and specifically with China is a very complicated question. The whole idea behind ITAR was to try and protect US intellectual property and security. I think most people will agree that it did okay for both those areas, but with the price that we have actually accelerated development elsewhere. It turns out to be easier for [other countries] to develop it themselves," he says.

Continuing on the theme of encouraging international collaboration, van Zyl adds: "Humans want to help each other fix problems. So [ITAR] is an unnatural restriction of collaboration in that sense. I fully understand the national implications . . . but my sense is that people are trying to figure out how we can work with other countries, including China, [while protecting] the intellectual property of the US. It's a big deal, it's worth a lot of money, and nobody wants to spend a significant fraction of their resources if someone else is going to take it and utilize it for their own gain. But I think there is some movement to eventually include China, and I think that's the right thing to do."

Scott Pace of the National Space Council had some thoughts on the subject as well, though a bit more proscriptive, that also begin with robotic exploration.

"I see potential opportunities in scientific cooperation such as unmanned scientific missions. For example, there's a commercial Chinese payload that's going to the ISS on NanoRacks, which I think is fine. There's scientific cooperation and data sharing in a limited way, which is similar to what we did with the Soviet Union in the Cold War . . . but manned spaceflight is about more than technology. It's a symbol. We didn't cooperate and bring Russians to the International Space Station until after the Soviet Union fell. Starting today, would we engage in human spaceflight cooperation with Russia given where we are? The answer is no. We did it in an earlier, more hopeful time and we did it because the political conditions had changed. So we don't do cooperation to change political conditions; political conditions change and then space cooperation follows that—it responds to and symbolizes that."

For the near term, ITAR restrictions remain firmly in place, and China will continue expanding spaceflight efforts on its own with both robotic and human programs. Most of these expansions will be of a peaceful nature, focused primarily on science. National pride and technological achievement will continue to be a driver for years to come, as was the case for the US and USSR during the space race. Still other programs will be driven by national security and military needs.

Looking out ten to twenty years at the possible relationships between the China National Space Administration and other space programs, Bruce Pittman has a more positive view. Pittman is a chief engineer at NASA's Space Portal at the Ames Research Center in California, and a senior executive of the National Space Society.

"So many people have sacrificed so much to get us where we are today," he

Chinese space suit for the Shenzhou program. Image credit: Rod Pyle

says. "Our future in space could be truly amazing, but we have some tough choices to make. I hope we can listen to our better angels and not to our fear or doubts. There are clearly some national security threats, and there's some militarization occurring in space, which is becoming more of a concern. One thing we can do to move the conversation forward is to concentrate on the positives and figure out how to make the development of space work well for everybody."

Pittman closed by suggesting that the political turmoil of today may lead to discussions that will initiate cooperative programs toward a better tomorrow. This will be, he thinks, in the context of a larger discussion about internationalism. "What are our obligations to each other? What is our government's obligation to us, and what do we owe to our government, to our country, to other nations, and to our species? These are all important questions that were implicit before, but at this point need to be addressed explicitly. We need to start challenging some of the assumptions that we've had for decades. We need to be better, and we will."

China's ascendancy in technology, and specifically in space, will make it increasingly difficult to continue excluding the nation from US and other space efforts around the globe. China is simply too valuable a potential partner to ignore. However, the reverse is also true. Despite its strong technical acumen, China would benefit from international cooperation, with the US the strongest possible partner. A solid working relationship between all the major players in space exploration will open the door for even more countries to join the effort, resulting in a truly global space program.

A Shenzhou spacecraft on display at the Shanghai Science and Technollogy Museum. It is an enlarged and modernized update of the Russian Soyuz. Image credit: Rod Pyle

Image credit: James Vaughan

CHAPTER 14

TRUCK STOPS IN SPACE

Infrastructure is not a very exciting word. In terms of an advertising executive's priorities, it's got absolutely none of the flash and dazzle of a rocket launch or the first boots on Mars. But when discussing the new age of spaceflight, and the new commercial opportunities that are arising, it is everything. Infrastructure will make the development and settlement of space affordable and routine. It's the only serious path forward. That's why many people in the space trade talk about it with such passion. Infrastructure is what it's all about.

So what is space infrastructure, exactly? Perhaps a metaphor is called for. Space infrastructure is much like the services that make your daily life livable in modern society. When you arise in the morning, you flip a light switch—the electricity powering that is part of infrastructure. Taking a shower? Plumbed water is infrastructure. Back the car out of the garage to drive to work? Roads and freeways are infrastructure. You get the idea. The gas you stop for, the drive-through Starbucks, and the internet that sends you the 113 emails you plow through when you arrive at work—all are a part of the infra-structure of daily life.

Of course, the specific elements of space infrastructure that will be needed depend on how we proceed. We might first construct orbiting fuel depots, or perhaps orbiting bases for construction of gigantic structures. Private companies are planning to build orbiting hotels. Industry and university groups in both the US and China are studying solar-power satellites that could provide energy for operations in space as well as for uses on Earth. Some deep-space mission concepts will require way stations for assembly of vessels bound for Mars and beyond. The fleets of reusable rockets being built by SpaceX, and soon ULA and Blue Origin, are also parts of space infrastructure. These are just a few examples.

Space is such an immature industry, the idea of satisfactory infrastructure is comparatively basic. Instead of the freshly paved freeway, we'll take a two-lane highway. We still need fuel, but we'll settle for regular instead of premium. No snacks or five-hour energy drinks necessary—just give us the essentials to make this endeavor thrive. In essence, we need the equivalent of basic office space, hospitals, hotels, gas stations, railyards, and truck stops in Earth orbit and beyond. In the new space age, if we wish to proceed beyond the expeditionary model of human space exploration—short trips to recon-noiter other worlds and then a return

home—robust infrastructure will be cru-cial. If we wish to expand the reach of our species beyond Earth, to live and work in space, infrastructure is, after affordable launch capabilities, the next critical milestone in the settlement of space. And for Space 2.0, the participants building this infrastructure hope that it will return them a profit over time, critical to private investment.

The National Space Society initiated its Space Settlement Summit in 2017, and it is now an annual event. In atten-dance the first year were a number of the top thinkers imagining and implement-ing Space 2.0, including private entre-preneurs, leaders within NASA, military officials, and individuals from the invest-ment community. They were there not just to discuss space *settlements*—human outposts in space—but also issues related to the infrastructure needed to enable space *settlement*—the development of space in many forms, both robotic and human, for global benefit. The two are inextricably linked.

Like so many subjects in Space 2.0, it can be easier to envision the distant goals rather than the immediate ones. We want fuel depots in space, resource extraction to provide that fuel, manufacturing using other in-situ resources, outposts, way sta-tions, communities, and so much more. But the first steps to achieving these goals are the most vexing.

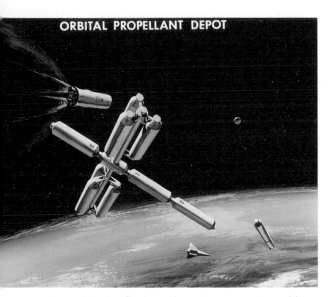

An orbital refueling depot as envisioned by NASA in 1971. This structure was intended to store fuel launched from Earth; the idea of collecting resources from the moon and asteroids was not yet widely accepted. Image credit: NASA

The US's final expedition beyond low Earth orbit: Apollo 17 in 1972. Gene Cernan driving the Lunar Roving Vehicle. Image credit: NASA

Humans have spent the last six decades making expeditionary dashes into space. There were those first forays into orbit . . . the leap to the moon . . . then almost fifty years of orbiting the Earth, in both stations and spacecraft. But none of this is real infrastructure. The space shuttle had only limited reusability. Even the ISS is an interim step and only usable via continuous resupply from Earth. The ultimate goal of space infrastructure is the ongoing availability of the assets and resources needed to live and work in space, derived from space-based sources—water and building materials from the moon and asteroids, for example.

One primary takeaway from the Space Settlement Summit was that no single answer can enable space infrastructure. There are a variety of plans and ideas, and we must collect and agree upon the best available subset of them and chart a path forward.

NASA contractor and NSS executive Bruce Pittman summarized the importance of infrastructure early in the proceedings: "The idea is to start a dialog regarding how to work within the solar

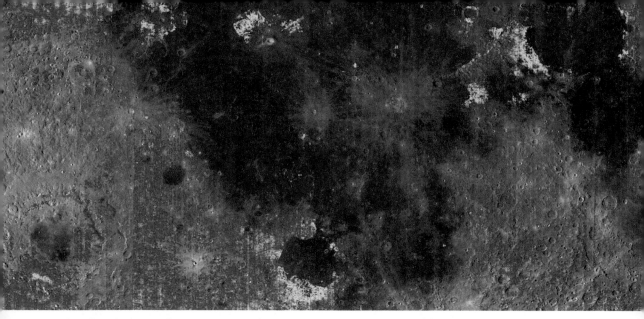

The distribution of water ice on the moon. This ice is a valuable resource for the creation of breathable air, potable water, rocket fuel, and more. Image credit: NASA/ Brown University

system for the next fifty years. It will take more than rockets, so let's broaden the discussion. We need to talk about the deep-space economy. We know how to make money in GEO—geosynchronous Earth orbit how can we close the business case beyond this?"

This statement framed an important question facing the entire space community, but in particular, the nonscientists. I say this because we can always formulate a rationale to go into space for *science*—NASA, ESA, and Russia's Roscosmos have been doing this for decades. But space science has always been supported by government and taxpayers, and is not inherently profit driven, though the spin-offs of the technological development it yields are tangible. *Business* in

space will elevate economic returns measurably. Indeed, the telecommunications industry alone has made many billions in orbit, but to move humanity deeper into space and allow it to stay, a solid business model is critical—the term "deep-space economy" says it all.

In the earliest stages, this means using the resources that are easiest to reach in the simplest ways. Water from lunar soil and possible ice deposits on the moon, as well as water in asteroids, can potentially be utilized for fuel, potable water, breathable air, and rocket fuel. Martian ice deposits and its atmosphere contain the necessary constituents for the same commodities. Lunar, asteroid, and Martian soil can all be mined to create bricks, concrete, and 3-D printed structures. Metal, glass, and

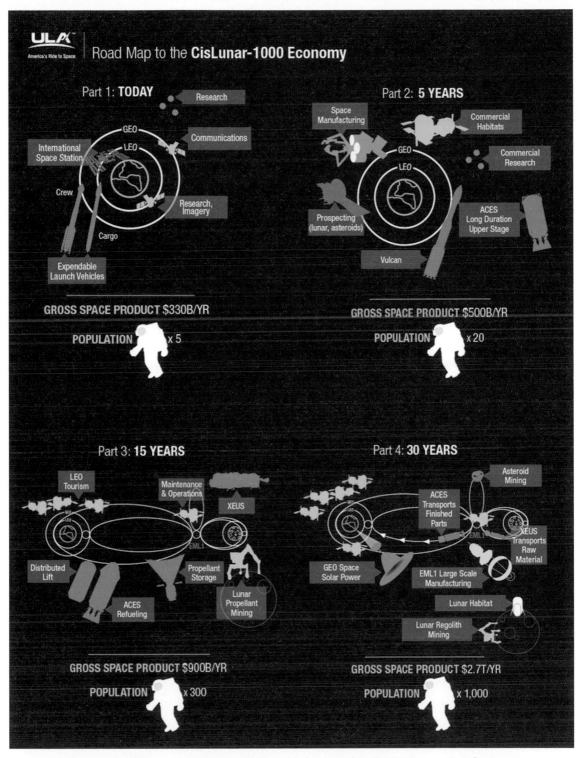

United Launch Alliance's CisLunar-1000, which sees the Gross Space Product at almost $1 trillion per year by 2030 or so. Image credit: ULA

other elements can be extracted from each of these bodies as well. Ongoing work at the ISS and research facilities all over the world has demonstrated that edible plants can be grown and produced in weightless conditions, and on other planets, with the proper techniques. With sufficient stocks of seeds and some enhanced sources of proper nutrition carried along, food should not present a major problem.

So, if we envision a time in which we have enabled the development of these basic supplies, we can consider the next step. Pittman moved on to discuss a future where these resources have been extracted and stored in fuel depots. "If I've got a surplus of goods, I can start selling those to other people," he explained. "That's how you move the frontier forward, and the process continues. What we want to know now is how to accomplish this at the lowest cost and with the most flexibility."

Let's get more specific about what this really means. Not surprisingly, a lot of smart people have been thinking about various parts of large-scale space infrastructure for decades at NASA, at aerospace companies, and at universities. But funding for Space 2.0 is limited, and new initiatives on the scale of the Apollo lunar landing program are unlikely. Expensive taxpayer-supported space shuttles are a thing of the past, and we are unlikely to build another $150

billion space station with tax dollars. Sprawling infrastructure programs fall in the same category—government funding alone will not be enough. A new model must be found to facilitate the building of this infrastructure. While creating early infrastructure will doubtless depend on some NASA money, likely via business partnerships such as NASA's commercial agreements to resupply the ISS, private investment will ultimately drive this process forward.

An example of this kind of partnership was presented at the conference by George Sowers, a former ULA vice president and currently a professor at the Colorado School of Mines. The plan he discussed is called the CisLunar-1000, ULA's initiative to have 1,000 people living and working in space by 2045 in a self-sustaining economy. The plan centers on the ACES space tug, which you might recall from Chapter 9. ACES could be used to transport supplies—air, water, fuel, and more—from wherever it is mined in space to where it's most needed, and then stored in depots. Some of these depots will be in Earth orbit, others out by the moon, and some eventually near and on Mars.

"There are [about] ten billion metric tons of water on lunar poles. We can start with refueling services," Sowers said. Water from lunar sources would be made into fuel, then stored. "This fuel availability provides for trade routes using ACES and XEUS." XEUS is a robotic lunar lander that will be able to transport useful supplies from the lunar surface to the storage depots.

While space entrepreneurs study ways to extract, transport, and store these resources, this may ultimately prove to be too risky for even the billionaires without government partnerships. Lori Garver, the former deputy administrator for NASA, feels that government partnerships with private industry are a good way to encourage sector growth. "I'm a huge believer in democracy and capitalism, and expanding those into space is a fabulous idea."

Some may see a contradiction in terms here—why should the government and, ultimately, taxpayers fund, or co-fund, these efforts, only to bring corporations profits in the future? Garver answers this question elegantly: "As we have seen in our capitalist society, the government invests in things that are hard, factors out some of the risk, and allows the private sector to move in and open up new markets. This will make spaceflight more competitive and means that we are going to space as a civilization as a fair and democratic society. So for me, a space program of the future just keeps expanding the envelope while private sectors come in and build behind it, and sometimes jump a little ahead, but they are in fact symbiotic." And, she continues, when the private concerns expand this infrastructure, the economic benefits ultimately come back to the nation that funded them, as was the case with railroads in the nineteenth century and airlines in the twentieth.

Our experiences on Earth have demonstrated that competition between entities, commercial or otherwise, drives innovation and growth. Space enterprise will be the same—competition between companies, large and small, will drive affordable access to reaching, living, and working in space, and will ultimately benefit national, then global, economies. Despite the billions invested by people such as Musk and Bezos, government will still have a large role to play, both in the US and internationally.

NASA represents the forefront of this kind of public-private partnership. No other national space agency has been faced with the rise of an entrepreneurial sector at the levels that have occurred in the US. As we have seen, this has already resulted in innovative and rewarding collaborations. But there's a sweet spot to be found in the relationship between NASA, the traditional aerospace companies, and the entrepreneurs. Identifying the correct blend will be one of the most profound challenges to NASA, and to the governments of other spacefaring nations, over the next decade or two.

The implication is that NASA would continue to transition from being a "we handle all major US space science and

Artist's concept of a Dragon 2 spacecraft bringing astronauts and cargo to the International Space Station under contract to NASA. Image credit: James Vaughan

human spaceflight endeavors" agency to a "we work with you in space by doing the hard stuff first and investing in the private sector to do the rest" organization. NASA has always hired outside contractors—the traditional aerospace companies—to build most of its space hardware. Until the twenty-first century, this was primarily done with "cost-plus" contracts—contractors are paid for their expenses fulfilling the contract, then paid an additional sum to allow them a profit.

In the past decade, NASA has shifted some of its procurement away from this model and toward what is called "fixed-price contracting," along with other similar agreements, to work in improved ways with SpaceX and Boeing to fly astronauts to the ISS. The goal is to share risk

and reward, and to induce contractors to invest more of their own resources in the development of new space technology. This allows NASA to contract out routine jobs, such as operating the ISS, launching rockets, and, eventually, establishing depots for space-derived fuels. NASA can then pursue the more exotic science- and exploration-oriented deep-space missions—both robotic and crewed—it has been so successful with in the past.

The idea that NASA should help to underwrite the development of space infrastructure with cooperative agreements has wide support. The question, then, is how to best achieve this kind of collaboration. What would such NASA-led infrastructure look like? How are NASA's dollars spent versus private investment? Where do the NASA missions stop and the private, entrepreneurial ones begin? Who benefits and how?

Bill Gerstenmaier, Associate Administrator for Human Exploration and Operations for NASA, suggests a plan for blending NASA assets and those of private industry. "If somebody will build a lander, I've got the infrastructure. If I have a habitation capability, such as an Orion capsule, I can then potentially use private-sector cargo transportation to and from a facility on the moon. I've got all the infrastructure pieces that will enable someone else, for just a cost of a lander, to have a lunar surface capability. Then we would team with them . . . to get to and from the moon. This allows NASA to gain more experience on the surface of the moon. So I don't have to pay for that right up front; that's covered by the interests of another party that wants to do things there. That's our general approach."[119]

In this model, NASA supplies the parts of a venture that it has already developed, or that make sense for it to develop, and private industry fills the gaps under cooperative public-private partnerships, eventually on its own dime. The growth of the communication satellite market provides some helpful examples here. The earliest versions flew in 1958 under NASA sponsorship. Then, in the 1960s, more satellites were sent into orbit, contracted by NASA and built by private contractors. In 1962, Telstar was the first privately launched communication satellite, a co-venture between AT&T, Bell Laboratories, the British and French national postal services, and NASA. Many others have followed.

An example of how this might evolve with regard to space infrastructure would be NASA providing the rockets to get to the moon, while private industry provides the landers and surface machinery for mining and processing resources there. This is exactly what Blue Origin is proposing with its Blue Moon lander, and Silicon Valley startup Moon Express with its lunar mining robots. The eventual expansion of such

Artist's concept of a 3-D printed structure being fabricated from lunar soil. Image credit: NASA

partnerships could see the mining of lunar ores and construction of structures with refined lunar material, with human occupants supported by moon-mined water and oxygen.

Since 1985, pioneering astronaut Buzz Aldrin has spoken and written widely about his ideas for space transportation infrastructure. He envisions spacecraft that follow permanent orbits between the Earth and Mars, to significantly reduce the cost and complexity of sending large numbers of people and cargo to that planet. These spacecraft are called Aldrin Cyclers.

The primary goal in Aldrin's concepts is the permanent settlement of space by humans. He foresees a logical progression of capabilities for sustaining people on the moon and then Mars, increasingly distant enclave enabled by its predecessors, beginning with new laboratories in low Earth orbit. Evolutionary design is a key feature, as are the use of resources on the moon, artificial gravity, and a combination of high-efficiency and high-thrust transportation systems. International participation will be important. These approaches enable a robust and affordable way of moving people from the Earth to Mars: cyclers. [120]

In Aldrin's designs, these cyclers use the gravity assist provided by Mars as they swing past the planet to return to Earth. Periodic trajectory corrections would be delivered by a propulsion system, either from chemical rocket engines or modern solar-electric engines, which use sunlight to power lower-output but longer-duration thrust. In either case, most of the flight is a "free ride."

The beauty of the cycler concept is that the large mass of the vehicle only needs to be launched and assembled once, and then it continues on its endless journey between worlds as long as it functions. Small shuttles would make runs from Earth to the cycler, then from the cycler to Mars as the cycler swings past its various destinations. Cyclers would be capable of carrying the life support equipment needed to sustain the crew. Even the requisite massive radiation shielding would no longer be an issue—materials from Earth, or found in space, need only be sourced once, then flown indefinitely.

Larger cyclers could include centrifuges that would provide a low-gravity environment for the crew, which would help them to maintain health during extended spaceflight. Cyclers would fly in pairs, with one transiting the outbound leg while a second paralleled that course in reverse, traveling from Mars back to Earth, concurrently.

Even if cyclers like the ones Aldrin envisions are implemented, more basic orbital systems will likely precede them. In October 2016, a working group met at investor Steve Jurvetson's offices in Mountain View, California, to discuss the creation of that sort of infrastructure. In attendance were aerospace engineers, scientists, entrepreneurs, and space advocacy representatives. Vast amounts of preparatory work had been done before the meeting—most of the attendees have been thinking about the infrastructure problem for at least a decade or two; some longer. Their combined knowledge and experience was crucial to creating a workable plan.

The meeting's mission was to create a five- or six-page document to send to the incoming presidential administration, after the inauguration in January 2017. The document would lay out a clear course to create additional public-private partnerships between NASA and industry, and ultimately toward developing a self-sustaining space infrastructure. The title was "Recommendations to the Next Administration Regarding Commercial Space."[121]

This document began with a suggestion to reestablish the National Space Council, which was enacted soon after the inauguration. As mentioned previously, the National Space Council is a body that reports directly to the president, via the vice president, on matters of space planning and policy. The group provides a set of guidelines, near and far term, for NASA and other US government activities connected to space.

There have been two iterations of the National Space Council since the founding of NASA in 1958. The first was the National Aeronautics and Space Council, which operated from 1958 to 1973, expiring about the same time as the end of the Apollo program. The NASA administrator, members of the defense

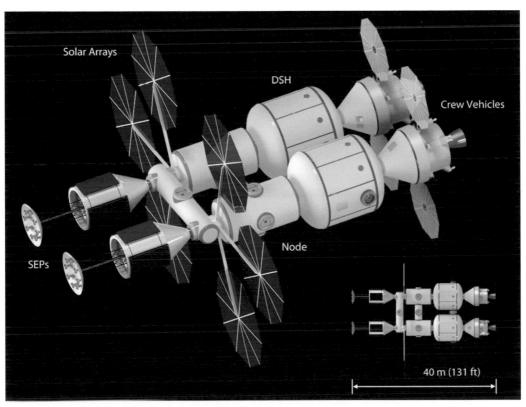

An Aldrin Mars Cycler design would traverse a permanent orbit between Earth and Mars. Similar designs could be applied for transport to the moon. Image credit: NASA/Buzz Aldrin

establishment, the Atomic Energy Commission, and others were included. This iteration of the council proved instrumental in establishing the Apollo lunar landings as a national priority.[122]

The council's second iteration operated from 1989 to 1993, during the George H. W. Bush administration. It was chaired by Vice President Dan Quayle and included the secretaries of State, Treasury, Defense, Commerce, and Transportation, the director of the Office of Management and Budget, the director of the CIA, the president's chief of staff, his assistant for National Security Affairs, the assistant for Science and Technology, and the NASA administrator. This council recommended the extension of funding for the threatened Landsat Earth satellite program, and worked on the Space Exploration Initiative, a 1980s plan to return to the moon permanently. This plan did not survive the transition to the Clinton administration.[123]

The current National Space Council is led by Vice President Mike Pence, with Scott Pace as the executive secretary, and includes the secretaries of State, Defense, Commerce, Transportation, and Homeland Security, the director of National Intelligence, the director of the Office of Budget and Management, the national security advisor, the chairman of the Joint Chiefs of Staff, and the NASA administrator.

The new National Space Council has access to the president, understands the complexities of dealing with the private sector, and is able to develop a long-range course that combines the best of NASA's innovation and exploration capabilities with strong support for infrastructure that will encourage private investment in space. Domestic security, economic growth, and military concerns centered on space are also of primary interest to this group.

Central to "Recommendations to the Next Administration Regarding Commercial Space" is the establishment of space infrastructure. As the paper put it:

For the exploration, development and eventual settlement of space to be truly sustainable, there must be a viable space economy to support it. **We recommend that the U.S. Government establish that one of NASA's goals should be to facilitate and promote a thriving space economy** [*boldface in original*].

The recommendations go on to discuss the global market for satellites and associated revenue, then they move to the larger financial possibilities in the space marketplace.

In 2015, United Launch Alliance (ULA) presented their "CisLunar-1000" view of the potential *for space development growth over the next 30 years. Their estimate was that the space economy could expand from its current $330 billion to $2.7 trillion by 2045. To make this projection a reality, the U.S. Government will need to play a vital but different role than it has traditionally fulfilled. The use of public/private partnerships as exemplified by the NASA Commercial Orbital Transportation Services (COTS) program and its use of funded Space Act Agreements (SAA) must become the norm instead of the exception. By aligning public and private strategic goals, dramatic financial leverage can be developed. A 2011 analysis of the development cost of the Space Exploration Technologies (SpaceX) Falcon 9 launch vehicle that was developed for the NASA COTS program was conducted by the office of the NASA Deputy Associate Administrator for Policy using the NASA/Air Force Costing Methodology (NAFCOM) computer modeling tool. This analysis showed an **almost 10X cost reduction** using the funded SAAs that were utilized by COTS as compared to the normal NASA cost plus contracts that are typically signed ($400 million for actual SpaceX Falcon 9 development vs $3.977 billion cost predicted by NAFCOM under a cost plus contract scenario).*

Multiple Falcon 9 first stages under construction at SpaceX. The company has able to leverage NASA contract revenue to help underwrite the development of some of its technology. This has resulted in a savings to the taxpayer of up to 90 percent over traditional aerospace contracting. Image credit: SpaceX

Next up was the value of NASA's commitment to private space companies when it comes to securing investment.

*Another key government initiative supporting commercialization of space was the Commercial Resupply Services (CRS) program. While COTS and the SAAs were utilized to demonstrate the capability to delivery cargo to the ISS, the CRS was a fixed price procurement contract for the actual delivery of payloads to the ISS over multiple years. **The CRS contracts that were awarded to both COTS winners*** *in an open competition allowed the two companies (SpaceX and Orbital ATK) to raise the money required to pay for their significant share of the COTS development costs.*

This section concludes with specific suggestions for an actionable pathway to this sort of privatization: "NASA should produce a plan to transition the ISS National Laboratory from the ISS to leased space in commercial LEO stations, and to assist new space businesses that use the ISS in a similar transition."

The idea here is to move the research operations currently taking place on the ISS to one or more private space stations, allowing NASA to mothball or deorbit the current ISS. This will spur investments in private stations, as well as reducing or eliminating the cost of maintaining an increasingly old and maintenance-intensive ISS.

The plan continues: "NASA should purchase rocket fuel and oxygen/water to use at any location in space (LEO, GEO, BEO [beyond Earth orbit]) from commercial entities if such commodities are commercially available. This policy will encourage the nascent asteroid and lunar mining industries, as well as lower the cost of an eventual journey to Mars."

The assurance that NASA will buy supplies from the companies now struggling to fund their programs to mine asteroids, build orbital depots, and construct space tugs to move these commodities will help them to secure needed investment.

It adds: "NASA bases/gateways/stations in any location in space, including the lunar surface, lunar orbit, and others, should contract with commercial services to provide cargo and crew to such stations. This policy will enable the development of economic and reusable cislunar transportation, and will support . . . an ultimate journey to Mars."

Paying commercial providers to transport both humans and supplies to NASA-built and -operated space stations and surface outposts will stimulate development, innovation, and competition in launch and transport service providers, speeding the creation of infrastructure while also keeping it cost effective.

The final recommendations outline NASA-backed programs to encourage breakthrough research from outside entities—companies, universities, and individuals—to advance goals in space.

Establish a Major Breakthrough Space R&D Program. *Throughout its history NASA has always been associated with major technological advancements, from the Saturn V that took American astronauts to the Moon, to the remarkably versatile, reusable but complex space shuttle,* *to the International Space Station that has been permanently occupied for [over] 16 years. No technological challenge seemed to be too great during this period. But recently, NASA's technological reach has been significantly reduced, and very few breakthrough technologies and/or capabilities now emerge from the agency's far more conservative and fiscally constrained endeavors.*

We recommend that the U.S. Government enable NASA to return to its cutting edge technology roots by establishing a significant ($1 billion/year) Breakthrough Technology R&D program *focused on providing the new capabilities and dramatic cost reductions to the aerospace industry that have been achieved in almost all other industries. Commercial companies, often backed by significant venture capital investments, are increasingly leading in the development of the cutting edge technologies required by our 21st century space program. NASA needs to team with these companies to encourage and mature selected technologies that can best enable ambitious future NASA missions. The establishment of an innovative and long term Breakthrough Technology R&D program, one that focuses on high risk but high payoff technology development and demonstration, would help not only*

NASA, but commercial space suppliers and users as well.

The paper concludes with specific areas in which this research might be conducted, such as the manufacturing of habitats, spacecraft, and parts in space; the construction of space solar power plants to beam clean energy back to our planet; and more.

Jeff Greason, a seasoned Space 2.0 executive, engineer, and now consultant, sees value in all the proposals, but especially the idea of an assured customer in space: "Anything that improves the demand for and reduces prices of space transportation is obviously helpful. Anything that builds up commodity markets for these goods and services that allows you to start putting together a business community is good. If propellant 'on orbit' has a price, then you can start to ask as a private enterprise 'Can I make money by putting a mining activity on the moon?' or 'Can I make money by mining near Earth objects?'"[124]

A number of groups are developing technologies critical to space infrastructure, in particular in low Earth orbit and on the moon. These range from small university-based teams to companies composed of seasoned entrepreneurs. As we've seen, companies such as Deep Space Industries and Planetary Resources are aiming at the mining of asteroids and the sale of the valuable resources found both in space (water and metals) and back on Earth (precious metals and rare earth elements, which are used in semiconductor manufacturing). Other companies, such as Moon Express, are developing landers that can go to the moon and prospect for water ice and other commodities for transport to orbital depots or use on the lunar surface. Blue Origin's Blue Moon lander will be capable of delivering the needed hardware to the moon and shuttling processed lunar resources back to Earth orbit, where they could be stored in depots.

There is one more part of the puzzle that needs to be worked out: government regulation.

Once you start conducting business anywhere, government regulation follows. This is necessary to establish and maintain an even playing field, but how much regulation, and of what type, is open for debate. Some will be critical to the success of free enterprise in space—the tricky part will be to establish the proper amount. In the past decade, government rules regarding spaceflight have increasingly fallen behind progress in the US private sector. One of the first tasks the new National Space Council set before itself was the updating of these rules, as requested by SpaceX, Blue Origin, and others.

For decades, the primary government-approved "space operating system" has been the previously mentioned Outer Space Treaty of 1967. To recap, it operates

much like the Antarctic Treaty of 1959: In the text, the Outer Space Treaty even states that a primary goal is to prevent "a new form of colonial competition." In practical terms, it means:

- No placement of weapons of mass destruction in orbit, in space, or on other bodies;
- No military maneuvers or tests in space;
- **No territorial claims on moons *or* other bodies.**

Therein lies the rub. The Space 2.0 companies interested in mining and resource extraction, whether from the moon, Mars, or asteroids, want some control over their workspaces—if not ownership or some kind of miner's claim or "stake," then, at the very least, a promise of noninterference. As Rick Tumlinson, longtime space entrepreneur and cofounder of Deep Space Industries, says: "We need a regulatory environment in space. Investors need to know that when you harvest resources, you can own them, sell them, and so forth." He adds that this will need to be a commonsense regulatory environment, one of resource ownership as opposed to property rights. "One precedent is fishing," Tumlinson explains. "When my boat is over the fish, I don't own them. I only own them when they are on the deck of my boat. I don't claim

A layer of water ice on Mars is seen in this enhanced image taken from orbit. Governments will have to cooperatively decide how such space resources can be utilized to be profitable. Image credit: NASA/Caltech/US Geological Survey

property rights on the ocean, just the right to fish."[125]

In 2015, President Obama signed the Commercial Space Launch Competitiveness Act. This was an effort to establish rights for American citizens to mine space resources in general accord with the Outer Space Treaty. The act also offered some clarity on how one might conduct operations in space without being challenged on legal grounds.

Jeff Greason reflects on this: "I think people will look back on that as a very significant policy action by the United States that is being echoed by other nations." Luxembourg has already made moves to

back a similar framework. Greason continues by noting that there was never any international discord over ownership of the moon rocks brought back by the Apollo program and Soviet robotic probes.

However, the Apollo moon rocks were scientific samples, gathered in another era. In the future, when space resources become commoditized and therefore have a recognized cash value, attitudes will likely change. International agreements will have to be reached regarding control of the resources that will fuel space infrastructure.

Control, if not outright ownership, of these resources will probably become one of the top concerns of the companies extracting, processing, and storing them. Greason speculates, "If one could acquire title to portions of other planetary bodies . . . those shares might be sold and traded on the market in some manner that makes those investments liquid." He adds, "People will want to buy more and more of those shares, because that creates more underlying commodities that they can speculate upon."

This notion goes beyond owning the resources mined off-Earth to property ownership, and that is a more complex discussion. But Greason feels that resource rights are the first step in the process: "The idea that you have the right to extract the resource from this celestial body and keep whatever you extract may point the way toward useful answers."

Chad Anderson of Space Angels says that regulation typically falls behind innovation and will take time to catch up. "One key challenge is that things are happening very quickly now, and this regulation didn't exist before because you don't build the regulations when there's nothing there to regulate."

Anderson continues: "Regulation is best handled when it is necessary and required . . . When airlines started their development, one of the first things they did was include the FAA in their plans, working with them early on to develop a regulatory structure to allow them to fly. As a result of that, the US has laws in place that allow airlines to operate. Now companies that are working to land on the moon have worked with regulators. It's important for companies to be forward thinking."

Anderson is referring to Moon Express, which in July 2016 received FAA approval of its plan to launch a mining reconnaissance probe to the moon. NASA and the US State Department also participated in the extensive audit of the mission plans. The FAA reviewed the launch path through airspace and the atmosphere up into orbit, and the State Department and NASA considered issues surrounding the landing of a private spacecraft on the moon and any associated implications regarding the Outer Space Treaty. This was an important step—not only for

Moon Express in its lunar ambitions, but for private spaceflight in general. This federal review demonstrated that the process of interacting with potential resources on other worlds can pass muster with the US government.[126]

Of note, the new National Space Council addressed the question of regulation in space during its first meeting in late 2017. "The idea here is to remove impediments," Vice President Pence said, referring to the streamlining and revision of existing laws to expedite rapid and competitive commercial space development. Pence also instructed the secretaries of Commerce and Transportation, as well as the Office of Management and Budget, to conduct a "full review of our regulatory framework for commercial space," with an eye toward simplifying regulation. Devising and implementing proactive new regulations to define and protect commercial operations in space is an important next step.[127]

As commercial activities in space increase and take a place alongside our already-extant government-supplied capabilities—which are substantial—many planners think that an enhanced means to protect them from foreign threats is essential. A number of national powers, China and Russia prominent among them, have taken steps to develop space-capable weapon systems, as has the United States.

On June 18, 2018, President Trump made an announcement that would create a sixth branch of the US armed services, an addition to the Army, Navy, Marines, Coast Guard, and Air Force, called the Space Force. The US military has been involved in space since the beginning of the space race, and its ongoing role is being looked at closely as the pace of development increases. The US military space budget is already larger than NASA's, at about $22 billion for "open book" expenditures; how much more is spent on classified operations is unknown. Up through 2018, most of this money was spent by the US Air Force Space Command, a part of the Air Force, founded in 1982 during the Reagan administration, but this may soon change.

Much of the country's investment in a Space Force will be directed toward perceived international threats in Earth orbit. Those stemming from Russia and China will receive particular attention, as these are the two international powers that, along with the US, are the most space capable. There has also been a role suggested for the US military in protecting American private industry resources in space. What form that protection will take, and in what time frame it will be implemented, is less certain. What is certain is that safeguarding space-based assets is essential, and it is likely that all the spacefaring parties will see this in a similar light. Security is just one more aspect of the essential infrastructure necessary for our expansion into space.

While *infrastructure* may not be an exciting word, it is certainly a thrilling prospect: The establishment of space-based resource extraction, transportation, and storage facilities near the Earth, moon, and eventually Mars will open the solar system to humans in ways previously seen only in science fiction. This will result not just in expanded opportunities to explore and develop space, but also in real benefits for people on the ground. Hundreds of thousands of jobs worldwide will need to be filled to support these endeavors, and the returns on investment in space will begin to take off in a big way once real infrastructure has begun to lower costs and increase opportunities. It's time to make this a reality.

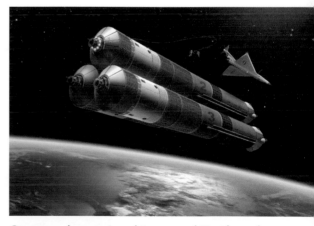

Storage depots in orbit around Earth and the moon will provide access to fuel and supplies needed to continue to other destinations in the solar system. Image credit: James Vaughan

Image credit: NASA

CHAPTER 15

DEFENDING EARTH

There's one more important component of our involvement with space that bears examination, and that is survival. Not just of you or me, but of the species, of everyone you know and love, and every other human being from the Arctic Circle to Tierra del Fuego. Because at any time, day or night, an Earth-destroying asteroid or rogue comet could appear on the radar and, once it impacts, kill every scrap of higher life on this planet.

This sounds like a plot from a Hollywood blockbuster movie, and it has been, more than once. While some of these stories were better told than others, at their core is one irrefutable fact: Earth could suffer a cataclysmic collision with an asteroid or comet. It's happened before. The biggest one we know of was the asteroid that was responsible for wiping out the dinosaurs (and, more positively, allowing the tree shrews—our ancestors—to evolve into us). An asteroid called the Chicxulub impactor struck Earth at the end of the Cretaceous Period; the impact led to the beginning of the Paleogene era.[128] The asteroid (or comet—it could have been either) was between six and nine miles across and struck near Latin America, off the east coast of today's Mexico, sixty-six million years ago. The result was the eventual extinction of up to 75 percent of plant and animal life on our planet. Earth was encircled with an unbroken layer of clouds and dust that caused an eighteen-month continual night. Temperatures plummeted, while ash and detritus rained for weeks on end. Neither plant nor animal were immune to its effects. It was a truly cataclysmic event.

Fast-forward to the present. We can see the remains of this crater—all one hundred miles of it—near the Yucatan Peninsula. Realize that at the time it struck, the planet was lightly populated by animals and vegetation, with no human activity such as agriculture or animal husbandry supplying food. Today, with 7.5 billion people, all of whom are dependent on growing food and raising animals for sustenance, the results of a similar impact would be potentially fatal for most of humanity. A few well-prepared groups might survive such a catastrophe, but most of our species would die off. People would be unable to grow food in the darkness and, without crops, would be unable to feed farm animals. Fish would die in the chilled oceans, eliminating another food source. It's a dark vision—and one with significant potential to come true.

NASA has mapped more than 1,400 asteroids of five hundred feet in

NASA is developing systems to provide early detection of dangerous asteroids, but more work will be required to divert them from impacting Earth. Image credit: NASA

diameter or larger that pose a serious threat to Earth's ecosystem. While these are not multimile-wide monsters, they are still city killers and would disrupt climate patterns for months. It's estimated that double that number may exist in orbits that threaten Earth, and that tally does not include unseen rogue rocks that could come from farther out in space. Comets, which are balls of ice and dust as opposed to large, solid rocks, also pack quite a punch. NASA thinks that it has identified about 95 percent of the threats over a half-mile wide, so there's some relief in that knowledge. But all it would take is one uncharted object, or a collision in space that would change a known asteroid's trajectory, to cause an imminent threat. In 2028, for example, asteroid 2001 WN$_5$ will pass well inside

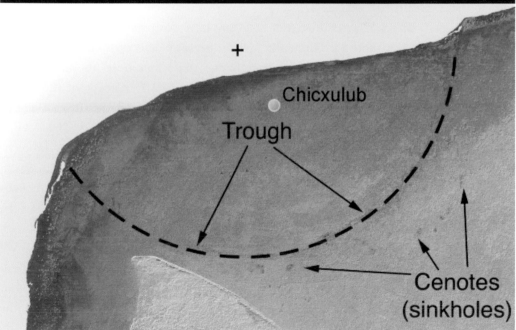

The site of the Chicxulub impact about sixty-six million years ago that is thought to have wiped out the dinosaurs and many other life forms. Image credit: NASA

the moon's orbit, at a distance of about 154,000 miles from Earth. While this may seem like a comfortable margin, anything crossing between lunar orbit and Earth gets the attention of asteroid watchers. At the speeds and distances that asteroids travel, it would not take a huge amount of deflection to set WN$_5$ on a collision course with Earth. At an estimated 2,500 to 4,000 feet in width, 2001 WN$_5$ is not a planet killer, but an impact would cause significant damage to the immediate region and affect global weather systems for decades.[129]

During the Obama administration, funding for the identification and tracking of these threats, as well as creating projections of their trajectories, increased by a factor of ten. But there are still many potentially dangerous objects that we have not charted. Even when we have spotted them and know their trajectories, that's not the same as eliminating the peril they present. The knowledge simply lets us know that the threat exists—we need to learn how to stop those threats from hitting the planet.[130]

We have done little to protect ourselves from such a devastating event and must learn how to better identify, track, and redirect these dangerous objects. Even with our enhanced awareness, we cannot spot them all with existing instrumentation. Recall the sixty-foot-wide chunk of space rock that exploded above the Russian city of Chelyabinsk in February 2013. Roughly 1,500 people were injured by the resulting five-hundred-kiloton explosion high in the atmosphere; the blast released about thirty times the power of the Hiroshima atomic bomb. The giant rock was not detected before it entered the atmosphere, so there was no warning. Some notice could have spared most of the victims from injury.

Lori Garver was deputy administrator at NASA when the large funding increases were allocated for asteroid detection, and she still feels it isn't nearly enough. "I think we are spending way too little on planetary defense, and we should be doing more of that science. We don't have, and never have had, very much money going into asteroid detection, and we need to change that. You need to find as close as possible to 100 percent of the bodies that will have a devastating impact on our planet and the people on it. That's when you can really start investing in the technologies to do something about it."[131]

Doing something about it means either the deflection or destruction of the threatening body. This is a huge engineering problem, and an expensive one. We have ideas and theories about how such interdiction might be accomplished, but they all require prior warning, and the most promising methods would require years—if not decades—of advance knowledge to be applied.[132]

Let's say NASA spots a half-mile-wide asteroid hurtling toward Earth today. We have no immediate response capability and watch helplessly as it grows ever closer to our planet's orbit. As it nears, we see that it is actually one large body with two smaller rocks orbiting it—such bodies are much more common in the solar system than we have previously thought. The main body has a mass of about 1.3 billion tons, the smaller moonlets much less. As the countdown continues we cross our fingers, holding out hope for a near-miss instead of a strike.

But luck is not with us this day, and the asteroid hits Earth. The main body impacts, with an explosive yield of 600,000 megatons, or about 12,000 times the power of the USSR's Tsar Bomba, the largest nuclear weapon ever tested. It slams into the Pacific Ocean, sparing the Earth's land masses, but this does not eliminate the danger. As the main asteroid careens through the atmosphere, it is moving so fast that it leaves a tunnel behind it—the air cannot collapse fast enough. Material that is superheated by the impact, including ocean steam and vaporized rock from the seafloor, travel back up that tunnel into the air. This material travels in a ballistic arc, almost into space, then falls back to Earth, still flaming hot. Meanwhile, the crater dug out of the ocean floor yawns about twelve miles wide. Enormous volumes of seawater now rush in to fill the impact hole and to replace the

Artist's concept of a large asteroid impacting in the ocean. Image credit: Solarseven | Dreamstime.com

mass of water that has been vaporized. The shock of the impact registers well over 9.0 on the Richter scale and may cause some seismic shifting as well.

A 140-foot-tall wall of water rushes outward from the impact site, headed for all the coastlines ringing the Pacific Ocean. Smaller tsunamis follow from the steam blasts after the main impact. All Pacific islands and many miles of coastal regions are scoured clean.

At the same time, one of the moonlets falls into the ocean, but the other has been sufficiently diverted to impact on land. Between the effects of the main asteroid's impact and the dry-land impact of the moonlet, dust and fragments have been blown into the atmosphere and join the ash produced by fires that have been ignited across the hemisphere by the initial rain of red-hot debris. A months-long period of almost nighttime darkness follows, dropping temperatures dramatically. This temperature shift, along with prolonged darkness, wipes out most remaining plant life, as well as killing plankton in the oceans. A greenhouse effect follows, gradually causing temperatures to return to previous levels . . . then accelerate well beyond, into the crop- and people-killing range (if any remain).

Acid rain falls for months, wiping out any plant life that was spared from the initial effects. This also destroys much of Earth's ozone layer and kills any remaining life in the shallow regions of the global oceans. The caustic acids cause respiratory harm to most animals and humans who are not protected. And that's all from an asteroid just over 2,600 feet wide. There are far larger ones lurking out there in space.

The foregoing is from a study conducted in 1989 and updated in 2013.[133] Different studies yield slightly different results and levels and types of damage, but this gives you a general picture—the effects of an asteroid strike would be catastrophic. The odds of this occurring are calculated in various ways. Some suggest the average time between such large impacts to be about 300,000 years. But it will happen someday, and anyone who has ever bet on a horse race knows the fallibility of depending on the odds. The

The DART mission will send an asteroid impactor to the moonlet of an asteroid called 65803 Didymos in the early 2020s. Image credit: NASA

risks are too large to ignore, for this danger is an inevitability. And some of those studies conclude that we are overdue.

What can be done to prevent such a global catastrophe? A number of mediating techniques have been studied, all of which require space infrastructure we do not currently have in place. Large rockets, along with the technology to deflect the asteroid, would be needed. Improved early detection equipment—both on Earth and in space—is required, along with the funds to cover operational costs. Test flights to harmless asteroids need to be conducted, and the results of attempted intervention monitored for their effectiveness.

While not well funded, such missions have been heavily studied, and a test project is moving forward. The DART program—it stands for Double Asteroid Redirection Test—is currently under development by NASA. The selected target is a moonlet orbiting an asteroid called 65803 Didymos. The moonlet, informally called Didymoon, is about 530 feet across. This smaller body is thought to be a good candidate for the test.

The DART spacecraft is a kinetic impactor, which will slam itself into the moonlet at a velocity of about 13,500 miles per hour. The energy of this impact should shift the orbit of the smaller body by about 1 percent, enough to be measured by Earth-based telescopes. DART is scheduled to launch in 2020 or 2021, with the test impact taking place in 2022.[134]

DART will provide a great start, but it is obviously just the beginning; DART is a tiny spacecraft being used to ever-so-slightly alter the trajectory of a tiny asteroid in a test. If it works, DART will give us a proof of concept needed to build much larger systems that could redirect bigger threats, but of course, these threats need to be spotted long before they approach Earth in order to be deflected from our planet. Think of it like billiards—the farther you are from the pocket when you deflect the ball, the wider the divergence when the ball arrives.

However we eventually proceed with planetary defense, it will not be an inexpensive undertaking, and, like many aspects of Space 2.0, it is probably best accomplished via an international consortium of concerned nations.

Infrastructure is clearly a key component of future planetary defense efforts. Larger kinetic impactors can be constructed of anything massive that can be aimed and propelled; lunar rubble would be a good choice, as would a properly prepared small asteroid. These masses could be parked in a safe spot—one of the Lagrange points, for example—then outfitted with propulsion units that would remain in standby. We would then be prepared to deflect the more common objects that threaten Earth. Having these assets established in space would be a huge plus. But such defensive spacecraft would be far larger than anything we could launch on a single rocket, and it will take vision, commitment, and money to prepare such a defensive capability.

Artist's concept of the destruction of an asteroid by nuclear weapons. Image credit: NASA

For those of you who are fans of world-smashing, apocalyptic asteroid impact movies such as 1998's *Armageddon*, the effectiveness of nuclear weapons against smaller asteroids has also been studied. Using nuclear warheads to redirect or destroy the target is a tempting idea, but the scientific community disagrees on the effectiveness of this gambit. Some think that a nuclear blast should succeed in diverting smaller asteroids, while critics fear that it would simply shatter the target, leaving a cloud of about the same mass hurtling toward Earth, only in smaller pieces. If this is the result of the blast, the resulting environmental havoc would remain catastrophic. Still others think it is unlikely to work at all.

Russia is currently considering one project of this kind, wherein they would repurpose an existing ICBM with a nuclear warhead to intercept and destroy an asteroid. The proposed test target is called Apophis, which is similar in size to Didymoon and will pass near to the earth in 2036. While we are currently tracking up to ten thousand potentially dangerous large asteroids, it is estimated that up to one hundred times as many smaller, but still dangerous, bodies—about the size of Apophis and Didymoon—pose a threat to our planet. The ability of an ICBM to launch quickly is appealing. Since they are weapons, they take far less time to prepare than conventional rockets.

Those tasked with planetary defense will be studying the results of the Russian test carefully.

Another deflection technology, called Laser Bees, is the subject of research by the Planetary Society and Scotland's University of Strathclyde. A swarm of small laser-carrying craft would be dispatched to the threatening asteroid well in advance of its predicted arrival at Earth. Firing in unison, the lasers would focus on a carefully selected part of the space rock, heating it sufficiently to cause a small amount of the surface to vaporize. Over time, this vaporized gas would act like a miniature rocket engine, creating enough thrust to deflect the asteroid away from Earth. A test mission is not scheduled, but it is another concept that warrants investigation.[135]

All the technology employed for protection need not serve just a single purpose, and if we're smart in how we approach the challenge, it won't. The same tools and infrastructure used for locating and prospecting asteroids can frequently be transformed for use in identifying and eliminating natural threats from space. Of course we get that mining tech developed faster by government partnering with the companies angling to extract resources from the asteroids, so potentially improved security—an ever-enhanced protective network on and above Earth—can be chalked up as another welcome side effect to those efforts at collaboration. And when we must invest in specialized equipment and systems to keep our planet safe, we will just have to find a way to do it. What's the alternative? In the end, we need a commitment to Earth's safety so that we may look to the heavens with anticipation of a bright future, rather than in fear. Planetary defense is a logical, necessary, and achievable part of the larger infrastructure of Space 2.0.

Firing a powerful laser at an asteroid would vaporize part of the rock, creating a propulsive force and potentially redirecting it away from Earth. Image credit: NASA

Image credit: James Vaughan

CHAPTER 16

SETTLING THE FINAL FRONTIER

*"Superhuman effort isn't worth a
damn unless it achieves results."*
—Ernest Shackleton, Polar Explorer

The Final Frontier. This phrase, used to describe space, became a staple of science fiction with the introduction of *Star Trek* in 1966. Used in relation to space settlement, it applies exquisitely. As the last of the unsettled frontiers, space tantalizes us with its potential. Settling this wilderness will not be easy, of course. As the late Paul Spudis, a scientist at the Lunar and Planetary Institute in Houston, said in a 2017 essay on lunar settlement, we will need to "arrive, survive, and thrive."[136] None of those things will be easy. But space challenges us to expand outward, and we will.

When I say "challenges us," I use that phrase deliberately. As discussed throughout this book, pursuing our goals in space with current technology is tough, expensive, and sometimes dangerous. But the more we apply ourselves to those challenges, the less of an impediment they will be.

The human settlement of space is the logical next step after affordable launch and the establishment of infrastructure for a number of reasons. To recap:

- Exploration: While there are still frontiers on Earth, we have barely begun to explore our own solar system.
- Commerce: It's still early in private and commercial spaceflight, but financial analysts see a bright and profitable future in space-based commerce. Off-earth mining, space tourism, and the vast range of services available from commercial satellites are just a few.
- Clean, Limitless Energy: Space-based solar power—stations that collect energy from the sun and transmit it down to locations on Earth—has the potential to provide more than enough power for every person on our planet.
- Resources: Though most resources mined in space, whether from asteroids, the moon, or elsewhere, are destined to be used in space, some, such as precious metals and rare earth elements, may be worth shipping home due to scarcity on Earth.

Let's look more broadly at space resources. Though our planet has managed to support a population that has now reached 7.5 billion people, we cannot continue to expand this number indefinitely. Earth's supply of natural resources is stressed, and the environment is increasingly polluted, despite our incremental efforts to mitigate the impact humans have on the natural world.

This subject is a source of considerable debate. Some will say that we are not even close to reaching resource limits, or that efforts to reduce polluting emissions have "fixed" the problem. The climate-change conversation is a topic that could fill an entire library. My point here is not to state specific, quantified figures, but to indicate a general trend: In the not too distant future we will outgrow our planet's *current* capability to sustain us without some major technological advances. Limitless clean energy, revolutions in food production, and the creation of clean fresh water may be provided via scientific breakthroughs, but there will come a time when expansion beyond Earth becomes a more practical or desirable answer.

Even if we can solve all our resource issues, the threat of calamity—man-made or natural—still looms large. Ensuring human survival by settling space is a concept that has been discussed for decades by some of our finest thinkers. It is a key driver of entrepreneurs like Musk and

Bezos—they feel that such expansion is imperative to preserve our species.

Physicist Stephen Hawking addressed this topic during a speech at Oxford University in 2016: "Although the chance of a disaster to planet Earth in a given year may be quite low, it adds up over time, and becomes a near certainty in the next 1,000 or 10,000 years . . . by that time we should have spread out into space, and to other stars, so a disaster on Earth would not mean the end of the human race."[137]

It is also important to understand that space settlement is not limited to *settlements* in space. According to Stan Rosen, NSS leader, career US Air Force officer, and a consultant to government and the space industry, "The settlement of space is far broader than colonies or outposts. It involves everything from the machines in orbit that make our lives better today, to enhanced robotics that will enable the betterment of human life on Earth as well as the expansion of humanity into space. Settlement includes space resource utilization, space fuel depots, research outposts, and more. It's about taking the best of our civilization and using new technologies and performing tasks in space that will revolutionize our civilization overall. So while human settlements may be a big part of it, space settlement is about much more than that. In broad terms, it's something that is happening today, as we speak."[138]

The earliest forms of space settlement include the utilization of satellites to change how we work and live. This GOES weather satellite, one of many, continuously transmits information about weather patterns over the US from a geostationary orbit. Image credit: NASA

But Rosen has one caveat regarding the future: "When you're talking about the trillion-dollar space economy, it will require more of the new economic model that we are just starting to use. The government is not going to spend a lot more on NASA, on national security, or on agencies like NOAA. Almost every major government organization benefits somehow from space, but they can't spend a whole lot more than they already are. These resources will be coming from the private sector, as they have been for some time, and this trend will continue to grow."

Current space-related economic growth is just the beginning. As the trillion-dollar space economy that Rosen and others predict becomes a reality, it will drive (and be driven by) our gradual movement into the solar system. This movement will offer humanity a far better chance for long-term survival. As Elon Musk put it in his 2016 talk at the 67th International Astronautical Congress in Guadalajara, Mexico: "I really think there are two fundamental paths [for humans]: one path is we stay on Earth forever, and some eventual extinction event wipes us out. The alternative is to become a space-faring civilization and a multi-planet species, which I hope you agree that is the right way to go."[139]

This is Musk's primary motivation for his efforts in spaceflight—he claims that the business model he has pursued, while profitable, exists primarily to support this goal.[140] There are far easier ways to make a fortune in the twenty-first century than building rockets. Witness the success of Jeff Bezos with Amazon, versus his ongoing investment in Blue Origin. But Bezos himself is bullish on space colonization for similar reasons. His stated desire is to have millions of people living and working in space. Speaking at the Smithsonian Institution in 2016, he said: "I wish there were a trillion humans in the solar system. Think how cool that would be. You'd have a thousand Einsteins at any given moment—and more. There would be so

much dynamism with all of that human intelligence. But you can't do that with the resources on Earth or the energy on Earth. So if you really want to see that kind of dynamic civilization as we expand through the solar system, you have to figure out how to safely move around and use resources that you get in space."[141]

Both these entrepreneurs are working for the eventual betterment of all. But just because settling space is a good idea does not mean it will come to pass. We have to be able to get people there in numbers, along with the infrastructure that will allow them to stay.

NASA studies from the 1970s were some of the earliest comprehensive work on space colonies. This vintage illustration shows a wheel-shaped colony that creates artificial gravity by spinning. Image credit: NASA/ Ames Research Center

A small lunar outpost constructed from Bigelow habitation modules. Image credit: NASA

Visions of extended human space settlement vary from person to person, group to group, and nation to nation. To some, it means huge space stations in Earth orbit, housing hundreds or thousands of people. To others, it means cities on the moon or Mars. To still others, it manifests as vast complexes in space, positioned at gravitationally stable Lagrange points surrounding the Earth and moon, with populations in the millions. Space settlement can also mean human enclaves on planets circling other stars.

Scott Pace of the National Space Council thinks that a permanent human presence in space will require a high degree of self-sufficiency.

"Space enthusiasts often assume that space settlement is a matter of manifest destiny," Pace says. "[But] that future will depend on whether or not we can really live beyond the Earth without resupply and whether or not we can do something that sustains us out there, or whether we are always dependent upon tax dollars from Earth. If the answer is we live off the land but we're always dependent on tax dollars from home, then space is Antarctica. We have a base, and maybe even tourists, but it is not a place that is self-sustaining. If the answer is we can do something economic out there, but we always have to come back to Earth for whatever reason, then it's like an oil platform. We go out, we work on the platform, do useful things, and then we come back. Those are radically different human futures in space and we actually don't know which one is right until we go out there and explore."[142]

In a nutshell, it's the expeditionary model versus the settlement model. In the former, we make runs out to places like the moon and Mars to explore, set up small outposts, utilize some local resources, and eventually return home. In the latter, we build bases in space and on other worlds, utilize the resources found there on an industrial scale, find ways to make those processes profitable, and settle the final frontier permanently.

NASA also supports the idea of extended, permanent space settlement, though you don't hear them talk about it often. In 2005, then-NASA Administrator

Future Mars bases will be sited underground or covered with Martian soil (as seen here) to protect the inhabitants from radiation. Image credit: NASA

Early outposts will likely be printed by machines using local resources. Humans can then populate them, expediting the process of settlement. Image credit: NASA

Mike Griffin addressed the topic in a *Washington Post* article. "[NASA's] goal isn't just scientific exploration. It's also about extending the range of human habitat out from Earth into the solar system as we go forward in time." He summarized his thoughts bluntly. "In the long run a single-planet species will not survive."

Griffin noted that entire species have been wiped out in mass extinctions on an average of every thirty million years. "If we humans want to survive for hundreds of thousands or millions of years, we must ultimately populate other planets. Today's technology is such that this is barely conceivable. We're in the infancy of it." He concluded by saying that one day "there will be more human beings who live off the Earth than on it."[143]

In 2016 a bill was introduced in the House of Representatives by Dana Rohrabacher, a congressman from California, that would make space settlement an official priority for NASA. It was called the Space Exploration, Development, and Settlement (SEDS) Act of 2016. The congressional summary states that the bill "requires the National Aeronautics and Space Administration (NASA) to encourage and support the development of permanent space settlements. Expanding permanent human presence beyond low Earth orbit in a way that enables human settlement and a thriving space economy shall be an objective of U.S. aeronautical and space activities. NASA shall obtain, produce, and provide information related to all issues important for the development of a thriving space economy and the establishment of human space settlements."[144] The bill was introduced in Congress but was not enacted. The subject will not fade away, though, as a handful of politicians and groups such as the National Space Society continue to push for its adoption.

Though NASA has been supportive of the notion of permanent space settlements, it has not set creating them as a specific goal. That may surprise you, but it shouldn't. The agency was created to lead civilian space efforts in 1958, during the Cold War and immediately following the USSR's success with Sputnik. The settlement of space was not on its radar at the time—matching and surpassing the Russians was the most immediate concern. Science and exploration, but not settlements, were key tenets.

Today, NASA is looking more carefully at enduring plans for space. Research and planning are ongoing for the technologies necessary to sustain humans beyond the Earth, both in orbiting stations and on the surfaces of other worlds. Research aimed at enabling extended survival in weightless conditions continues on the ISS. The Jet Propulsion Laboratory is working on at least two technologies to extract oxygen, and eventually water, from the Martian atmosphere. One of them, an experimental instrument called MOXIE (for Mars OXygen In-situ utilization Experiment), will fly aboard the Mars 2020 rover and test the technology shortly after landing. Other experiments at NASA field centers focus on using robots to gather soil on the moon or Mars, then extruding it as blocks and other construction components. NASA's robotic probes on the moon and Mars are cataloging water supplies in the form of large bodies of ice on and below the surface. NASA is also developing rovers, life support systems, and other technologies to enable long-term survival in harsh environments.

Once scientists and engineers develop the technologies required to support humans in off-world settlements, what might these outposts and colonies in

The Mars 2020 rover will carry an experiment called MOXIE that will attempt, for the first time, to convert a sample of Martian atmosphere into oxygen. Proving this technology is a first step toward extended human habitation. Image credit: NASA

space and on other worlds ultimately look like? Some of the first serious work on colonies in space was done by Gerard O'Neill of Princeton University. His 1977 book, *The High Frontier: Human Colonies in Space*,[145] laid out a daring blueprint for how humanity could inhabit the nearby bodies in the solar system using giant cylinders positioned at the Earth-moon Lagrange points. The cylinders would be constructed out of materials processed from lunar ores and shot from the moon's surface with a giant mass-driver gun. His plan was summarized in a paper published in 1974, in *Physics Today*:

"It is important to realize the enormous power of the space-colonization technique. If we begin to use it soon enough, and if we employ it wisely, at least five of the most serious problems now facing the world can be solved without recourse to repression: bringing every human being up to a living standard now enjoyed only by the most fortunate; protecting the biosphere from damage caused by transportation and industrial pollution; finding high quality living space for a world population that is doubling every thirty-five years; finding clean, practical energy sources; preventing overload of Earth's heat balance."[146]

O'Neill founded the Space Studies Institute to develop his ideas, and the resulting studies represented early contributions to serious thought about the large-scale, permanent settlement of space. His design for the structures at the heart of his plan was vast in scope: Each "colony" (he called them "islands" early on) would consist of two cylinders twenty miles in length by five miles in diameter. They would rotate along their long axes to provide artificial gravity. The cylinders would turn in opposing directions to counteract gyroscopic forces that might cause a single unit to drift from its sun-pointing orientation. The hulls would consist of six horizontal strips, alternating between solid metal for the living surfaces within, and transparent materials to allow sunlight to enter. Another, separate ring would be allocated for agriculture, leaving the remaining internal surface areas as residential and open-park zones. An industrial manufacturing region would be located toward the cylinder's center, where weightlessness would aid industrial processes taking place there.

The atmosphere inside these structures would be one-third that of sea level on Earth, about the same as is used in most human-rated spacecraft, and would include nitrogen to more closely approximate the air we are used to breathing. It was thought at the time the plans were released that the hull and the cylinder's atmosphere would provide adequate protection from space radiation, though more recent research suggests more would

A period illustration of O'Neill's space cylinders from the 1970s. These were intended to be many miles in length and powered by the sun. Image credit: NASA

need to be done on that front to keep the inhabitants safe.

The Space Studies Institute continues to promote these and other ideas about permanent colonies. Many other proposed designs have come and gone since O'Neill first published *The High Frontier.* The smaller Stanford Torus, about one mile in diameter, was a wheel-shaped, less expensive variation on O'Neill's design. [147] On the other extreme is the massive Dyson Sphere, theorized by physicist Freeman Dyson—a metallic globe large enough to hold a star at its center and contain a human population at a sufficient distance from the star to comfortably support their survival. [148] This is a highly theoretical idea

and is more of a thought experiment than a workable design, at least with any foreseeable technology.

Concerns have been raised about what kinds of governments might take hold in space settlements, and what possible risks they might face from ever more powerful economic and military establishments back on Earth. These questions are beyond the scope of this book but make for interesting reading. [149]

Today, we are on the eve of some truly inspiring prospects for early outposts in the final frontier, places that could become reality with the next two decades.

First and most timely is NASA's space Gateway (formerly the Deep Space Gateway and later called the Lunar Orbiting Platform-Gateway or LOP-G). This

One of NASA's designs for the Lunar Orbiting Platform-Gateway that is scheduled for the 2020s. Image credit: NASA

initiative reflects work performed across many years by NASA and numerous contractors and was announced by NASA in early 2017.[150] A preliminary agreement to jointly study the project as a cooperative goal has been signed between NASA and the Russian space agency, and a number of experienced aerospace contractors are developing ground-based prototypes.

The Gateway will be crewed by four to six astronauts in thirty-day rotations a few times per year and is similar to a smaller version of the ISS, except that it will be assembled in an orbit around the moon. The Gateway is modular in construction, like the ISS, and will probably be launched by the Space Launch System when that rocket becomes available.

The idea behind the Gateway is to provide a scientific research platform for NASA and its international partners beyond Earth orbit. The structure's location near the moon is intended to provide deep-space experience for longer-term voyages, such as a Mars mission. It should also facilitate a return to the lunar surface and provide a test bed for equipment and techniques for use on both the moon and Mars.

The Gateway is not necessarily a single-destination station, however. It will have solar-electric propulsion units that convert sunlight to low-thrust rocket propulsion, both to keep it in place once in orbit around the moon, and to move it as needed to other possible orbits. As Bill Gerstenmaier, the chief of human spaceflight at NASA, put it: "What's exciting is that with electric propulsion it doesn't have to stay in one orbit around the moon. It can be equatorial, polar, or halo [a larger, more distant orbit] to allow telerobotic activities." He continued, "This is not a science station. It is a research facility that can be moved to alternate locations with the power propulsion bus and which stays in orbit for roughly twenty to thirty years. It is also the transportation node potentially for going to Mars."[151]

The Gateway is a first step toward larger bases both in space and on the moon's surface. It may also prove to be a critical component of ESA's Moon Village, which should begin operations in the late 2020s as an international, possibly government/private consortium. Notably, China has expressed interest in participating. Several countries have expressed interest in participating, and early plans include 3-D printing of structures using moon-sourced resources, possibly prototyped by robots operated from the Gateway.

The purpose of the Moon Village, according to Johann-Dietrich Wörner, ESA's director general, would be to provide a permanent installation to support scientific research, lunar mining and other business enterprise, even lunar tourism. As planned, the facility embraces "the core

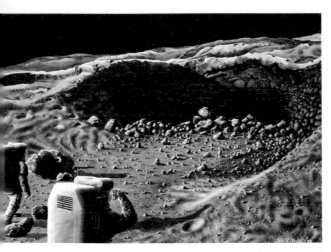

Part or all of the Moon Village could be sited inside lava tubes, such as the one pictured in this illustration. The rock overhead provides complete protection from radiation. Image credit: NASA

of the concept of a village: people working and living together in the same place—and this place would be on the moon . . . we would like to combine the capabilities of different spacefaring nations, with the help of robots and astronauts."[152]

The village's location and specific functions are open for discussion. "No human has ever visited the far side of the moon," Wörner notes. "Astronomers want to set up radio telescopes there because it is shaded from Earth's radio pollution. Building a telescope with innovative techniques like 3-D printing, perhaps using lunar soil, would enable us to look much deeper into the universe." The lunar pole regions have remained

similarly unexplored by humans, though unmanned missions have located ice there.[153]

Planners of the Moon Village are looking at a modular, expandable design, and expect to have between six and ten "pioneers" working there by 2030, and up to one hundred by 2040. They suggest that growth will come with the proof that resources can be harvested from the moon.

The confirmation of a large lava tube on the moon in 2017 may accelerate these plans.[154] Data from a Japanese lunar orbiting probe, when combined with measurements from a NASA orbiter, has confirmed suspicions that a large hole in the Marius Hills region in fact leads down to a cave large enough to contain

The Marius Hills lava tube on the moon—the interior is large enough to contain an entire outpost. Image credit: NASA

the city of Philadelphia. Siting an outpost in such a cavern would provide ample protection from micrometeorites, temperature extremes, and most importantly, radiation.[155]

Elon Musk recently expressed his support for a moon base as well. In mid-2017 he said that he felt that a lunar base would be the logical extension of the missions of the Apollo program, and would be inspiring to the public at large.

At the International Astronautical Congress in 2017, Musk mentioned that his newest rocket design would have lunar abilities, and he showed images of a possible moon base. His message was clear: Access to, and possibly the development of the moon is on SpaceX's agenda. His massive Super Heavy/Starship will be at the core of these plans.

The first modules of NASA's Gateway are planned to fly in the same general time frame as Musk's new rocket, and given the likelihood of time slips in both programs, they may well be operating in cislunar space concurrently. If SpaceX keeps to its overall development plan for the Super Heavy/Starship, we can expect to see a consolidation of efforts. And let's not forget that Blue Origin should be operating its heavy-lift New Glenn rocket and Blue Moon lander by the early 2020s. The competition between SpaceX and Blue Origin will augment the SLS and provide more affordable access to the moon. In

sum, a return to the moon looks likelier than it has since we last visited in 1972.

A human habitat on Mars, initially small and eventually expanding to contain hundreds or thousands of individuals is, after the moon, one of the next "big picture" items for human settlement in space. Along with NASA's ongoing research for smaller outposts, Musk envisions a sprawling Mars city in conjunction with his new rocket.[156]

The earliest Mars habitats will probably not be dissimilar to lunar ones. In either case, a large enough area is required to allow people to work, sleep, and play in relative comfort. This requirement is based on experience on long-term ISS deployments as well as studies of small populations in complexes such as Antarctic bases. The structures must of course be airtight and capable of protecting the inhabitants from radiation. The latter can be accomplished in a number of ways, the simplest of which is to bury part of all of the habitats under lunar or Martian soil, or to build these bases inside lava tubes, which appear to exist on Mars as well as the moon.

Regardless of which approach is followed, and who goes first, the moon and Mars are obvious initial targets for human expansion into the solar system. Both bodies have gravitational fields that should be beneficial to human health, and both contain raw materials useful for fueling the rockets that will travel there, as well as to build settlements and provide resources needed to keep people alive. Cislunar space is also viewed as an attractive environment for settlements.

Given what we know today, let's project a likely course of near-term developments in human spaceflight.

A PLAN FOR SPACE 2.0

Note that I said "plan." The spaceflight business is full of timelines, but timelines by definition have dates, and delivery dates for space accomplishments, especially in the speculative private sector, are rarely certain. When a technology is tested and proven, it will begin to be employed in regular use. This process is all part of Space 2.0, the opening of the solar system to a human presence. Rather than specifying exact dates, let's look at a possible progression across the next couple of decades.[157]

Remember, this is an aspirational timeline, and should be considered a best attempt, based on a combination of what companies and national entities are planning, tempered by finances, technical challenges, and politics.

SUBORBITAL FLIGHT: 2018–2022

- Space Tourism: Blue Origin and Virgin Galactic both operate suborbital flights for tourists and scientific payloads by 2020. A number of small companies provide suborbital launches

for short-duration robotic experiments.

EARTH ORBIT: 2018–2030

- Commercial Spaceflight to Orbit: By 2020, SpaceX and Boeing fly astronauts to the ISS. SpaceX and Blue Origin also execute private missions in 2019 or 2020, with SpaceX sending paying passengers to the moon by 2023. By 2020, SpaceX's Falcon Heavy and Blue Origin's New Glenn are competing on cost and efficiency. By 2022, SpaceX's Super Heavy/Starship completes testing and begins selected flights. The net effect: Costs of space access, for both human and nonhuman payloads, drop dramatically. Russia expedites plans to replace lost revenue from paid ISS flights from the US with a new spacecraft design and a new booster. Virgin Orbit, Stratolaunch, and a number of other providers deliver small satellites into orbit, lowering the cost of nonhuman payloads. A number of international launch providers, most of which are in China, introduce lower-cost launch services in an attempt to compete with US firms. The Chinese companies utilize reusable launch vehicles. India competes with lower labor costs and high efficiency, continuing to fly expendable rockets.

- Government Spaceflight to Orbit: Early in the 2020s, NASA's Orion, lofted by the SLS booster, flies shakedown missions to low Earth orbit and then to the moon, carrying components of the Gateway. China assembles its new modular low Earth orbit station from 2023 through 2025. Crew rotations follow, with some non-Chinese citizens eventually joining them. Russian crews continue to fly Soyuz, then its successor spacecraft to the ISS. China, India, Europe, and Russia strive to reduce launch costs to compete with private US launch providers.
- ISS Decommissioning or Privatization: The ISS is partially repurposed and partially deorbited in 2025. The Russians detach and reuse their modules, and others are temporarily mothballed in orbit for reuse by a consortium of private companies with NASA support.
- Private Space Stations: Companies such as Bigelow Aerospace, operating privately or in cooperative NASA partnerships, place habitats in low Earth orbit. These accommodate both corporate and government crews, along with well-heeled tourists. By 2030, humans have had a permanent presence in low Earth orbit for 30 years.

CISLUNAR SPACE: 2018–2030

- Cislunar Infrastructure: By 2022, ULA's Vulcan makes routine deliveries to orbit and flies the ACES reusable upper stage at every opportunity. These remain in orbit, and ULA works with Blue Origin to position orbital fueling systems to allow the ACES stages to be reused for repositioning payloads in Earth orbit, as well as transporting Blue Moon landers.
- Lunar Space Stations: The Gateway begins to take shape in lunar orbit in the early 2020s. This flies as a NASA-contracted partnership made up of SLS and/

or privately provided launches between 2021 and 2025. The Gateway starts as a modular assembly with components that parallel those of the ISS and is augmented over time by the use of Bigelow expandable modules.

- Lunar Landings—Robotic: Private companies self-fund and land robotic craft on the moon by 2020. The first landings are experimental; within two years these are replaced with second-generation landers to begin the first experiments in lunar mining and other resource extraction. By 2022 private firms commence the early stages of commercial ore processing and 3-D printing of components, as well as water extraction and storage. While most of this work is autonomous, by about 2023, complex tasks are overseen telerobotically from the LOP-G. China and India operate independent robotic processing plants.

- Lunar Landings—Crewed: NASA, ESA, and private industry work to put a crewed lander on the moon by 2024. SpaceX succeeds in the same year. The first components of a small research station are delivered as well. Between 2025 and 2030, China lands its first crew on the lunar surface by or soon after 2030.

- Lunar Base—Crewed: A lunar outpost—a collaboration between NASA, ESA, Russia, Blue Origin, SpaceX, and other private entities—begins to take shape in the late 2020s. The goals are life-science research, lunar exploration, and further advances in lunar resource extraction and utilization. China lands habitation modules on the moon. By 2030, work has started on infrastructure that will mean a permanent human settlement on the moon.

- Lunar Mining and Manufacturing—Robotic: By the mid-2020s, private companies complete tests in lunar resource extraction and ramp up to industrial scales. Blue Origin's Blue Moon lander has been in operation for a few years. Water extraction proceeds at the lunar poles, and surface processing of lunar soil proceeds near the equator. Though these sites are fully automated, they are visited periodically by small crews. Legal guidelines for control and ownership of material found in space have been established in US and international law; these guides prove critical to the ultimate success of space commerce.

- Pilot Space Settlement: Blue Origin or SpaceX launch elements of a privately funded space colony to low Earth orbit for testing in 2028. They are then boosted to a Lagrange point for further assembly and testing. Habitation takes place in the mid-2020s with a crew of twelve people, to expand to twenty-four within two years. The company adds modules each year via on-orbit additive manufacturing, and the population grows, tripling annually. The overarching goal, regardless of who ultimately launches and finances such a facility, is to have a permanent human presence in cislunar space. The colony uses some degree of artificial gravity, probably induced by a rotating structure, as it is highly desirable for human health. The ultimate self-sufficiency of such facilities is also a key goal. Reproduction off Earth is studied, to assure human suitability to live and thrive as a true spacefaring species.
- Earth Defense Systems: The world's government and private

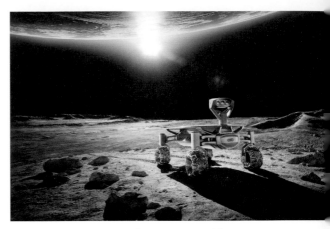

Artist's impression of a proposed lunar roving vehicle sponsored by automaker Audi. Image credit: ESA

space companies develop methods for the early detection and deflection of asteroids, comets, and large objects that might threaten Earth. Multiple solutions are found, both from targeted research and from the repurposing of tech meant for commercial functions, but much of the planetary defense system remains experimental.

- Space-Based Solar Power and Other Resource Development: Early work is undertaken to investigate the harnessing of solar power in space, both for use there and transmission back to Earth. Other in-situ resource development, possibly from asteroids and the moon, follows.

- Mars Orbit: A private-public collaboration accomplishes the delivery of a crewed platform into Martian orbit after assembly at the space Gateway, using Orion and SLS. Components arrive there in the early 2030s. China and India work with international partners on various robotic orbiting missions.

- Mars Landings—Robotic: By the mid-2020s, a significant number of landers and rovers, from multiple nations, join NASA's machines on Mars, working in coordination with an international fleet of orbiting spacecraft. SpaceX lands an unmanned Starship on Mars in 2022 in order to test flight and landing technologies and in-situ resource extraction. By 2026, SpaceX sends another Starship uncrewed to Mars, carrying a heavy payload of outpost infrastructure. This provides an opportunity to pre-position in-situ resource extraction technologies, with equipment provided by NASA, ESA, Russia, and possibly the private sector. Both India and China run robotic operations on Mars.

A robotic Mars Sample Return rocket blasts off, in this artist's impression, carrying Martian rock and soil samples back to Earth for examination in the late 2020s. Image credit: NASA/JPL

- Mars Landings—Crewed: Mid to late 2020s, SpaceX delivers the first passengers to Mars in conjunction with NASA. Or, NASA lands the first astronauts on Mars with the SLS/Orion system, augmented by a habitation/transit module. Or, SpaceX and NASA race to Mars. Or, SpaceX delays its ambitious plans for the Super Heavy/Starship and NASA lands humans on Mars in the 2030s. No sure bets on this goal, but it will be exciting to witness the first human landing on the Red Planet.

- Mars Base and Settlement: Between 2030 and 2040, work progresses toward a crewed Mars research station and, later,

Early arrivals on Mars trek to the Viking 1 landing site in Chryse Planitia to plant the US flag, in honor of the first successful probe to land on that world. Image credit: NASA

a human settlement. Humans make great progress toward a self-sufficient, thriving settlement on Mars.

The larger goal of human settlements in the solar system depends on the success or failure of the pathfinder programs outlined above. The process of establishing long-term bases may be undertaken by governments working with traditional aerospace contractors, or by private industry alone, but will most likely be achieved through a combination of both. National governments will continue to pursue space settlements because they think it is important in the long term, for a variety of both rational and emotional reasons. Corporations

will ultimately undertake such ventures because they know that there is potential profit in them. Extremely wealthy individuals may pilot such projects for philanthropic reasons, but nobody is wealthy enough to support an ongoing space settlement—yet. No matter the project's backers or their motivations, though, it's going to take individuals willing to venture out into the unknown to make any plan a reality.

Jim Keravala is CEO of OffWorld, a company developing robots that will extract and process space resources. He has a long history in the space business. In his view, the traditional motivations for individual migration and settlement on Earth—access to opportunities and resources—will not apply to much of the settlement of the moon or Mars, at least early on. Keravala believes that day-to-day life in these places will be very hard, more akin to working on an oil rig than the luxurious existence foreseen by people such as O'Neill.[158]

However, the opportunity to improve our circumstances in space lies just over the horizon, in Keravala's view. He sees much of the heavy lifting for settlements being done by machines and a few human overseers, but this will lead eventually to the construction of luxuries of which we can scarcely dream on Earth. "The requisite for a better life in space is to create an oasis . . . the only reason people will settle in space, post commerce-driven, is to achieve that better life."

Note that he said "post commerce-driven." This is an important point. Mining, resource extraction, and commercial services, such as telecommunications, will drive the first phase of space exploration, and it is only after this phase has set the infrastructure in place that those who are motivated primarily by the desire for "a better life" will find the final frontier appealing. "So far, there is no better place to settle than Earth," Keravala noted. "It's the baseline 1.0 definition of an oasis for now. To close the loop, we must take the issue of somewhere to live and

Rob Mueller (left) and Buzz Aldrin (right) look at a prototype lunar prospecting machine at a NASA event. Image credit: NASA

create luxury living, total creative freedom, abundant cuisine, longevity, unlimited entertainment, an ability to explore, and [a good system of] regulation."

Former NASA deputy administrator Lori Garver drilled down a bit further into human nature. "Over the long term, as you evolve space activities to benefit civilization, ultimately we have to go into space to assure our survival and to benefit civilization . . . Settling in new lands is what humans have always done, and I think we need to continue to do so, not only for resources, but for liberty and the human spirit. To be blunt, it is a combination of fear, greed, and glory. The will to survive is innate in every living thing. In general, people will want to go for their own reasons. These processes are not usually government driven. I'll add that the people who go on their own are typically the ones who stay."[159]

Jeff Greason emphasized property rights—the "greed" part of Garver's position. "If one could acquire title to portions of other planetary bodies, one could use the tried and true methods of past colonization efforts on Earth. Many such schemes have been tried in which the settlers get shares in the settlement. Up front, something like the Hudson Bay Company might put up the investment, and then offer the opportunity for people to come and work there—and subsidize them to go—but then [those settlers] have to work for some number of years in order to pay off the cost of their transportation and support."[160]

Former NASA astronaut Franklin Chang-Díaz thinks it is destiny, the oldest of trump cards that will drive settlement: "I believe it is our destiny to populate space. Space holds the key to our survival as a species. We need to prepare humanity to embark on that journey before the growing environmental and social stresses of our own planet extinguish our capacity to do so."[161]

Ultimately, whatever rationale one selects for seeking a future in space, those reasons must make sense to the stakeholders who count the most: private enterprise, government, and perhaps most importantly, *you*. Any efforts at space settlement must be supported by popular opinion—governments need it to survive, and private enterprise needs it for investment. As companies like SpaceX and Blue Origin push the boundaries of what can be accomplished, popular support for these ventures has been on the rise. The last major survey of public attitudes about spaceflight was conducted in the US in 2015 by the Pew Research Center.[162] It found that 58 percent of Americans felt that it was essential that the country remain a leader in space exploration. About 64 percent felt that spaceflight was a good investment for the country. Almost 70 percent held a favorable opinion of NASA. While another major survey has not been conducted since the launch of the Falcon Heavy, media accounts reported great excitement about the accomplishment worldwide.[163] Public support seems to be solid and appears to be increasing as people see the aspirations of Space 2.0 becoming achievements. This, along with the ever-increasing amounts of private investment in spaceflight, portend great things ahead.

Early Mars settlers work their way back to a transport vehicle to return to base as a dust storm approaches. Planetary exploration and settlement will be humanity's greatest adventure. Image credit: NASA

Image credit: James Vaughan

CHAPTER 17

A NEW AGE DAWNS

Like that day in 2011 when the final shuttle took to the skies, the morning of February 6, 2018, dawned bright and clear at the Kennedy Space Center. The mild winter weather reminded me why so many people move to that part of the country, and the crowds surrounding NASA's spaceport that day made it feel as if half the Midwest had done just that. But most of these people had traveled hundreds or thousands of miles not to relocate, but to witness a seminal event in the history of spaceflight: the inaugural launch of a new rocket, SpaceX's Falcon Heavy.

They might have been in for a long wait. As the countdown progressed toward the big event, the launch of the largest American rocket since the last Saturn V left the cape in 1973, a timely departure was far from certain. SpaceX was many years behind schedule with the Falcon Heavy, which had been originally slated to fly sometime in 2013. The challenges of combining three Falcon 9 rockets, the mainstay of SpaceX's rocket fleet, were far more substantial than even Elon Musk, the brash founder of the company, had envisioned. As he put it a few months earlier, "It actually ended up being way harder to do Falcon Heavy than we thought . . . Really way, way more difficult than we originally thought. We were pretty naive about that."[164] As it turned out, most of the rocket's central core had to be completely redesigned and strengthened to withstand the rigors of launch with two more Falcon 9s strapped to its sides.

The Falcon Heavy's redesign took five years longer than anticipated, but redesign it they did. And they tested and tweaked, and tested again, until Musk felt it was ready to fly. Even then, with all the possible ground tests completed, Musk publicly gave the first test flight a roughly fifty-fifty chance of success. But the only way to know for sure was to fly it, and the day for that test had come.

On February 6, 2018, the fruit of his labor stood ready on Pad 39A in

The Falcon Heavy awaits its first test flight the evening prior to launch. Image credit: SpaceX

Florida, the same launch facility from which the final Saturn V had lofted Skylab into orbit, back in 1973. The Falcon Heavy vented great white plumes of liquid oxygen into the midday air. It was a thrilling scene, watched by millions on broadcast TV and internet video. SpaceX supplied its own online channel to stream the event, complete with its own commentators.

The countdown ran smoothly, with none of the holds observers of space launches had become accustomed to in the era of the shuttle, when it was not unusual for delays to last hours or days. Mechanical issues, computer glitches, bad weather (the cape is notorious for its freakishly fickle weather), or even an errant pleasure boat that had strayed into the safety zone

The Falcon Heavy inside SpaceX's hangar in Forida. Image credit: SpaceX

off the coast could cause a launch delay. But today, none of those occurred, and the count continued as planned.

As a space author and journalist, I was invited to the press site for viewing. Launch parties were another option; there were many going on that day, as the aerospace community was brimming with excitement. So, too, were the millions of normal folks who follow spaceflight activities. But for this day, I chose to watch from home, because I wanted to experience the moment in privacy. I knew I would not be the only person to get misty-eyed if Musk were successful—there are countless others who grew up during the heady days of NASA's lunar landing missions who would feel similar emotions—but as the opportunity to watch the dawn of a new space age neared, I wished to experience it alone. I was not disappointed.

As the crowds assembled near NASA's new countdown clock to watch the digits tick down to T minus zero, the conversation was muted—everyone seemed to be holding their collective breath. Many were surprised that there had been no major holds. They were waiting to see if something would stop the clock at the last moment.

As the announcement of "T minus thirty seconds . . ." echoed across the viewing area, a cheer went up. At T minus eighteen, the launch director could be heard to say, "Launch director on countdown one, SpaceX Falcon Heavy, go for launch." Another cheer. Spouts of water could be seen on the video screen, bathing the launchpad to damp out the acoustic pounding it would soon experience as the twenty-seven rocket engines powering the behemoth roared to life.

Another announcement: "T minus fifteen, stand by for terminal count." The female voice then counted backward from ten, echoing countdowns from the earliest days of spaceflight. At about five seconds, gouts of steam could be seen spewing from the flame deflection troughs on Pad 39A. At T minus zero, orange flame was visible from the base of the rocket. Musk's engineering masterpiece was on its way.

Even though the Falcon Heavy has only about half the total thrust of the shuttle, it is far lighter and leaves the earth much more quickly. That morning, the rocket swiftly pounded into the sky, arcing to the east toward orbit. The terrific boom of the engines reached the crowd shortly thereafter, assaulting everyone's senses and bringing smiles to those 100,000 faces across the cape.

At two minutes and thirty-three seconds, the two side boosters separated from the center core. This too was not a sure thing. Musk is not a fan of explosive separation mechanisms—the way NASA and every other spacefaring nation has always accomplished booster separations—and insisted on using

The Falcon Heavy roared off Pad 39A after a smooth countdown, just one more stunning indicator that Space 2.0 has arrived. Image credit: SpaceX

pressurized nitrogen to blow the boosters free of the main rocket. But off they came without a hitch, as stunningly shown by the onboard cameras affixed to the central rocket.

The show was not yet over. As amazing as it was to see a private citizen build and launch one of the largest rockets ever created, Musk had one more treat in store. The core of his plans has always been to make his rockets reusable, therefore saving money and lowering launch costs. He had managed to land a number of his Falcon 9 boosters already and had even reused a few, proving both his engineering acumen and the value of his ideas about reusability—re-flying these recycled rockets had already saved his customers money. The Falcon Heavy was designed to do the same thing, but in this case, would require three boosters, all navigating from high in the upper atmosphere, to autonomously navigate back to Earth and land themselves. Would it work?

A bit after eight minutes into the flight, the two side boosters plummeted toward the landing zone at the cape, firing their rockets at the last moment, then settling down to the concrete landing pads, just as intended. Triple sonic booms from each booster, the result of their high-speed transonic reentries, rippled across the landscape. Again, the crowd went wild.

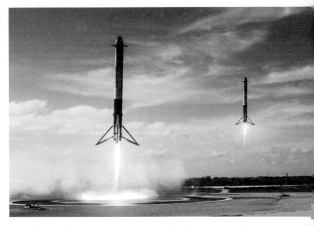

As the Falcon Heavy's payload entered Earth orbit, its two side boosters landed back at the cape, settling on to their assigned bull's-eyes as planned. Six sonic booms were heard due to the twin boosters' transonic reentry. Image credit: SpaceX

A minute passed and the video feed from the landing barge, many miles out to sea and awaiting the arrival of the central core booster, cut out. It was later discovered that the rocket's center booster ran low on fuel and slammed into the ocean at an estimated three hundred miles per hour, destroying it and damaging the landing barge. But this was an experimental flight after all, and this mishap was a small price to pay. Musk chalked it up to another learning experience.

High above the earth, the second stage continued firing, carrying the as-yet-unseen payload into orbit.

Then, perhaps the most remarkable moment of the flight from a cultural, if not an engineering, perspective occurred. In an inspired moment certain to go down in pop-culture history, as the rocket's fairings split into two halves and fell free, the onboard cameras revealed Musk's very own cherry-red Tesla roadster in orbit. Perfectly timed on the telecast, David Bowie's song "Life on Mars?" played in the background. It was a surreal moment that would stretch on for the next twelve hours, as live video of the car, with the earth turning slowly in the background, streamed around the globe. In the driver's seat sat a mannequin trussed up in a custom SpaceX pressure suit, with the left arm casually draped over the driver's door. Musk dubbed the helmeted figure "Starman." One could easily imagine a live person there, enjoying the view from a Tesla convertible as it slowly orbited Earth.

Hours later, the second-stage rocket engine fired again, propelling the car and its inert passenger beyond Earth orbit and toward the orbit of Mars. Past the Red Planet, the Tesla looped back towards Earth and then return to deep space in an endless oval. Musk has estimated that the Tesla could remain in that orbit for up to a billion years, being slowly eroded by the harsh radiation of interplanetary space. Of the launch itself, he said simply, "I guess it taught me crazy things can come true. Because I—I'd say I didn't really think this would work. Because when I see the rocket lift off, I see a thousand things that

Elon Musk's personal Tesla roadster being prepared for launch. Image credit: SpaceX

The moment of truth: the first Falcon Heavy lifts off right on schedule—a rarity for untried rockets. Image credit: SpaceX

could not work, and it's amazing when they do."[165]

A plaque on the dashboard reads, *Don't Panic*, a reference to Douglas Adams's best-selling 1979 science fiction classic, *The Hitchhiker's Guide to the Galaxy*, a Musk favorite. On a circuit board inside the car, another smaller message reads: *Made on Earth by humans*. It is a message to anyone who might happen upon the remarkable and potentially puzzling payload in the distant future.

Space 2.0 had finally, truly arrived. It had started years before, but the launch of the Falcon Heavy seemed to mark the resounding advent of this new space age. As I watched Starman sitting in his roadster, blue oceans and white clouds drifting distantly behind him, I recalled that final act of the shuttle program seven years earlier, an event infused with wistfulness over what could have been. This day represented a profound call to a new era.

Space 2.0 is upon us. Our adventure in the universe beyond our earth is just beginning.

Views of Starman

On February 6, 2018, SpaceX launched Elon Musk's red Tesla roadster as a test payload aboard the Falcon Heavy. The car orbited Earth before the upper stage engine was refired, sending it on a long, looping, and permanent trajectory past Mars's orbit and back to Earth. Image credit: SpaceX webcast

Image credit: James Vaughan

CHAPTER 18

YOUR PLACE IN SPACE 2.0

As late as the end of the 1990s, your options were limited if you wanted to be "involved" in spaceflight and associated fields of endeavor. If you were interested in space science or planetary science, you either worked for NASA or another space agency, or in academia. If you were an engineering type, you might work for NASA or an aerospace contractor or subcontractor. A handful of people made a living creating TV shows or writing about space in periodicals or books, but that was about it.

Today, the opportunities for involvement in space have expanded dramatically. There are seemingly as many paths to engagement as there are people who want to tread them, whether as a vocation or an avocation. Many still pursue careers as scientists, engineers, or managers in traditional aerospace companies or government agencies, but thousands of others are involved with smaller, entrepreneurial outfits, creating cubesats, spacecraft components, software, or any of the dozens of other kinds of technologies and tools that are necessary for the development of space. Many new operations aren't involved with space systems at all but are integral to the use of space-derived data to provide a multitude of high-value services, from agriculture to commercial satellite imaging. People ranging from college undergrads to retirees are working in their garages, from home offices, on kitchen tables, or in academic labs, on projects that will ultimately touch some aspect of humanity's expansion into the final frontier. There are even companies and associations working on space sports, space entertainment, and space tourism. New ideas enter the marketplace every week.

Other people are less directly but still meaningfully involved in space via volunteering for nonprofits. Their activities range from teaching kids about space science and engineering to meeting with senators and members of Congress about space-related issues. Still others participate directly in crowdsourced science. One example of this is a recent project that searched for new planets beyond the edge of our solar system by drawing on the efforts of over 60,000 online volunteers, each of whom scanned reams of imagery to help determine new places to look for planets. It's all incredibly exciting, and you can find a way to be engaged in Space 2.0 regardless of your trade, education level, or background.

The internet allows for motivated people to find each other in ways never before possible. You can find countless forums online where space science and space-related ventures are discussed and participants sought. In the past few years, NASA and other agencies and institutions have crowdsourced everything from astronomical research, as noted above, to spacecraft engineering and human spaceflight support. Hundreds of programmers and budding engineers participate in hackathons every year, designing new software and technologies for spaceflight; some of these contests are sponsored by NASA, others by universities and space advocacy groups. Anyone can get involved in these efforts—you don't need to be an engineer or scientist. The sponsoring groups need managers, media and outreach people, and people with various other skill sets to make the contests run smoothly. All you need is the drive to participate.

A number of nonprofit organizations promote citizen involvement with space science and exploration, and these are an excellent way to connect with space efforts. Membership in these groups also affords people a wonderful way to engage with hundreds of others who are interested in similar subjects. For many, space becomes a lifelong engagement.

Of these citizen groups, the National Space Society is one of the oldest and most highly structured—it's concerned primarily with human spaceflight and space settlement. It was begun in the 1970s by

Wernher von Braun of NASA, and in 1987 merged with the L5 Society, which was founded on the space settlement ideas of Gerard K. O'Neill at Princeton University. The Planetary Society, which focuses primarily on robotic exploration of the cosmos and, as its name suggests, planetary science, was founded in 1980 by Carl Sagan, Bruce Murray, and Louis Friedman, all of whom were major figures in planetary exploration from JPL. The Mars Society started in 1998, operates two Mars-simulation stations in Utah and the Arctic, and promotes the exploration of the Red Planet, as laid out by cofounder Robert Zubrin and others. These and a number of similar organizations are listed below.

Each of these organizations has a unique set of goals and activities. The NSS has one of the broadest, and its functions include well-organized lobbying to drive space activities in the US and internationally. Mark Hopkins, currently the chairman of the executive committee for the NSS, says this about NSS membership, activities, and plans for the future:

"Our goals include moving society into space. This involves far more than just astronauts and engineers. There are opportunities for engagement for just about everyone—lots of things which have little connection with space per se, but are critical to the pro-space movement. For example, people who are good at public relations are very valuable. People who can learn how to lobby Congress, people who are interested in writing articles, people who are interested in the activities of large nonprofit organizations, are all critical to our success. If you look at our membership, most of them are not directly involved in space—I would say maybe 10 percent of our membership actually have jobs in the space sector. There are far more NSS members involved in computers than there are in the aerospace industry, for example. No matter what your interest or specialty, there are plenty of ways to become involved with space through the National Space Society.

"This really is a cause," Hopkins continues. "We are not talking just about space, but about the future of the human race as a whole. So if you're interested in making a better future for humanity, or promoting a new American dream for all humanity, where each generation is better off than the previous one, then the NSS offers a great way to become directly involved in creating that future."[166]

To all my fellow space enthusiasts out there: We are no longer barred from participating in humanity's greatest adventure. The new space age has room for all of us to be engaged to the limits of our skills, our talents, and our passions. Now is the time to get involved, because space holds the key to our future on this planet—and beyond.

International participation is welcomed by most of the membership-based organizations, and some, such as the NSS, have a substantial network of international chapters. The depth of your involvement is limited only by your willingness to join and engage with these organizations. The following list highlights groups that are primarily oriented toward individual memberships; many more professional organizations and specialty groups can be found online.

SPACE SCIENCE, ADVOCACY, EDUCATION, AND ENGAGEMENT ORGANIZATIONS

NATIONAL SPACE SOCIETY (NSS)

The NSS is an independent pro-space organization built on a grassroots foundation. It publishes papers and a quarterly magazine called *Ad Astra*, holds multiple annual conferences, and is composed of separate chapters globally. It also engages in direct political action for space activities in Washington, DC, and engages in STEM/STEAM activities. Membership is open to all.

Address: PO Box 98106, Washington, DC, 20090-8106
Phone: (202) 429-1600
Email: nsshq@nss.org
Web: space.nss.org

THE PLANETARY SOCIETY

The Planetary Society is an educational organization, with the stated goal to "know the cosmos and our place in it." It publishes a quarterly magazine, organizes volunteer political activity, holds annual conferences, and produces a weekly radio program and podcast called "Planetary Radio." Membership is open to all.

Address: 60 South Los Robles Avenue, Pasadena, CA 91101
Phone: (626) 793-5100
Email: tps@planetary.org
Web: planetary.org

THE MARS SOCIETY

The Mars Society is focused on enabling and accelerating human exploration of Mars. It holds an annual conference, organizes various other events, and runs two Mars analog research stations—one in the Utah desert and another in the Arctic. Membership is open to all.

Address: 11111 West 8th Avenue, Unit A, Lakewood, CO 80215
Phone: (303) 980-0890
Email: info@marssociety.org
Web: marssociety.org

TAU ZERO FOUNDATION

The Tau Zero Foundation sponsors research into breakthrough propulsion and technologies, participates in annual conferences in interstellar spaceflight and near-term space and energy technologies, and publishes technical papers. Membership is open to all.

Email: info@tauzero.aero
Web: tauzero.aero

AMERICAN ASTRONAUTICAL SOCIETY (AAS)

The AAS strives to "strengthen and grow the space community," influence space policy, educate the public about space, and provide networking opportunities. The organization holds a variety of conferences each year and provides technical publishing opportunities. Membership is open to all.

Address: 6352 Rolling Mill Place, Suite 102, Springfield, VA 22152-2370
Phone: (703) 866-0020
Email: aas@astronautical.org
Web: astronautical.org

AMERICAN INSTITUTE FOR AERONAUTICS AND ASTRONAUTICS (AIAA)

The AIAA claims to be the world's largest technical society dedicated to the global aerospace profession. It organizes national meetings and conferences, publishes papers, conducts outreach programs, and advocates for aerospace professionals. Membership includes categories for academicians, researchers, students, corporations, and various specializations. It is intended for aerospace professionals, from their earliest student days through retirement, but events and conferences are open to the public.

Address: 12700 Sunrise Valley Drive, Suite 200, Reston, VA 20191-5807
Phone: (800) 639-2422
Email: custserv@aiaa.org
Web: aiaa.org

THE SPACE FOUNDATION

The Space Foundation is a group composed primarily of corporate entities. It houses a new space exhibition at its headquarters in Colorado Springs, Colorado, and hosts an annual space conference, the Space Symposium. Memberships are limited to corporations, but individuals can purchase a Discovery Center Passport that includes admission to the center and provides email updates.

Address: 4425 Arrowswest Drive, Colorado Springs, CO 80907
Phone: (719) 576-8000
Web: spacefoundation.org

INTERNATIONAL SPACE UNIVERSITY

The ISU is not an enthusiast group, but rather an internationally based educational consortium that specializes in space curricula. Applicants must have a bachelor's degree, and some professional experience is preferred. The ISU is a powerful way for young aerospace professionals and scientists to boost their careers.

Address: Parc d'Innovation, 1 Rue Jean-Dominique Cassini, 67400 Illkirch-Graffenstaden, France
Phone: +33 3 88 65 54 30
Email: info@isunet.edu
Web: isunet.edu

BRITISH INTERPLANETARY SOCIETY

The BIS is a UK-based pro-space and space education organization. It offers regular lectures, symposiums, and other activities and resources, including the monthly publication *Spaceflight* magazine. Membership is open to all.

Address: Arthur C. Clarke House, 27–29 South Lambeth Road, London, SW8 1SZ, UK

Phone: +44 (0) 20 7735 3160
Web: bis-space.com

SPACE FRONTIER FOUNDATION

Like the NSS, SFF supports the human settlement of space. It hosts an annual conference, is engaged in political activities with regard to space and spaceflight, and publishes a monthly newsletter. Membership is open to all.

Address: 9318 Warren Street, Silver Spring, MD, 20910
Phone: (623) 271-2045
Email: info@spacefrontier.org
Web: spacefrontier.org

THE SETI INSTITUTE

The SETI Institute supports the search for extraterrestrial intelligence through direct observation and other research. It also participates in astrobiology research. The Institute offers scholarships and research opportunities, as well as a wide array of public engagement programs. The newsletter is free, and membership is not required.

Address: 189 Bernardo Avenue, Suite 200, Mountain View, CA 94043
Phone: (650) 961-6633
Email: info@seti.org
Web: seti.org

SPACE STUDIES INSTITUTE

The SSI was founded by Gerard O'Neill in the 1970s and continues to pursue his dream of space settlement. SSI is particularly interested in deep-space colonies and lunar-resource utilization. It supports many types of research directly related to spaceflight and publishes technical papers and other materials. Membership is open to all.

Address: 16922 Airport Boulevard, #24, Mojave, CA 93501
Phone: (661) 750-2774
Email: admin@ssi.org
Web: ssi.org

STUDENTS FOR THE EXPLORATION AND DEVELOPMENT OF SPACE (SEDS)

SEDS is an organization of student chapters across the US and internationally. It helps students in science, engineering, and related academic programs to gain access to and attend professional conferences, publish their work, and develop their leadership skills and professional networks. SEDS hosts an annual conference and offers competitive scholarships. Most colleges and universities with aerospace engineering and science programs support SEDS chapters. Membership is also open to high school students, and students may join regardless of their technical background.

Address: 3840 E Robinson Road, Amherst, NY 14228
Phone: (704) 778-8648
Email: sam.albert@seds.org
Web: seds.org

CANADIAN SPACE SOCIETY

The CSS is a membership organization dedicated to furthering Canadian involvement in space efforts. It hosts an annual conference, publishes a newsletter, and promotes discourse through its chapter structure across Canada. Membership is open to all.

Address: 1115 Lodestar Road, North York, ON M3J 0G9, Canada
Email: president@css.ca
Web: css.ca

INDIAN SPACE SOCIETY

The ISS is an India-based pro-space organization dedicated to promoting growth in India's economic, technological, and political space efforts. It runs a number of educational and outreach activities at schools throughout India.

Phone: 91-8209244957
Email: indianspacesociety@gmail.com

Web: indianspacesociety.wixsite.com/
isswix

YURI'S NIGHT

Yuri's Night is an internationally active organization that promotes an annual celebration of achievements in space on a global scale. It hosts nearly three hundred events in fifty-six countries. Paid membership is not required, and by signing up online you will be kept apprised of the group's activities. Goals are to build space-oriented communities, enhance international cooperation, create industry leaders of tomorrow, and encourage scientific literacy.

Email: hello@yurisnight.net
Web: yurisnight.net

EXPLORE MARS

Explore Mars works to advance the human exploration of the Red Planet. It hosts an annual conference, publishes a newsletter, and organizes numerous educational and outreach activities internationally. It also participates in direct political action to further the goals of Mars exploration.

Address: PO Box 76360, Washington, DC, 20013
Phone: (617) 909-4425
Email: info@exploremars.org
Web: exploremars.org

AUSTRALIAN SPACE RESEARCH INSTITUTE

The ASRI is a volunteer-run group that supports Australian space activities with an emphasis on launch systems. It holds an annual conference and is membership based. Membership is open to all.

Address: PO Box 266, Mt Ommaney QLD 4074, Australia
Email: asri@asri.org.au
Web: asri.org.au

ICARUS INTERSTELLAR

Icarus Interstellar is a membership organization dedicated to advancing humanity's expansion beyond the solar system. The organization supports a network of college campus-based chapters. Icarus Interstellar runs a yearly conference and publishes research papers, as well as promoting education and outreach efforts about interstellar travel.

Email: info@icarusinterstellar.org
Web: icarusinterstellar.org

THE MARS INSTITUTE

The Mars Institute works to advance the scientific study, exploration, and public understanding of Mars and its moons, as well as the moon and near Earth objects. It publishes technical studies and provides public engagement opportunities. The institute also operates the Haughton-Mars

Project Research Station on Devon Island in the Arctic, a leading Mars analog field research facility.

Address: c/o Dr. Pascal Lee, NASA Ames Research Center, MS 245-3, Moffett Field, CA 94035-1000
Email: info@marsinstitute.net
Web: marsinstitute.no

SPACE GENERATION ADVISORY COUNCIL

The SGAC is an international organization that promotes the engagement of young people in space-related fields. It works closely with the United Nations Program on Space Applications. SGAC provides scholarships and outreach efforts as well as public information on educational and engagement programs.

Address: c/o European Space Policy Institute, Schwarzenbergplatz 6, 1030 Vienna, Austria

Phone: +43 1 718 11 18 30
Email: info@spacegeneration.org
Web: spacegeneration.org

Together, we will take humanity to the stars.

ACKNOWLEDGMENTS

This book involved well over one hundred individuals who gave generously of their time, effort, and in some cases hard-earned dollars to support its creation. My deepest thanks to all of you.

The next generation of space exploration, what I call Space 2.0, is a moving target. While writing this book, which took almost three years, each set of revisions involved not just editing and improving the manuscript, but updating content due to rapid changes in the spaceflight landscape. As the manuscript progressed, a new US president was inaugurated, NASA cancelled some programs and added new ones, the National Space Council was reactivated, and a dozen new promising, well-financed space start-ups appeared. Large private space efforts, such as those from SpaceX and Blue Origin, matured and evolved over the same period. New spacecraft flew, notable among them being the Falcon Heavy, which launched successfully in February 2018 and removed most remaining doubts about the viability of privately developed reusable rockets. It's been an amazing few years.

I would like to single out a few key individuals who contributed greatly to the publication of *Space 2.0*.

During the roughly five years from first conception to completion, my stalwart agent, John Willig of Literary Services, Inc., was encouraging and helpful throughout the process. While interest in the marketplace waxed and waned with regard to NewSpace, John continued to represent and promote the book, regardless of my own frustrations. Finding a partner in the publishing business who will work so hard and diligently, both with publishers and the author, is not easy. John has been that partner with unceasing good cheer, optimism, and words of support. He is a rare individual.

To the wonderful folks at BenBella Books, I extend my deepest thanks. Glenn Yeffeth took the leap of faith to engage what has turned out to be a mammoth project. Vy Tran, Lindsay Marshall, Sarah Avinger, Monica Lowry, Adrienne Lang, Alicia Kania, and Heather Butterfield all contributed much to the successful completion of the project. Finally, Jim Lowder, the fine editor assigned to this challenging project, was insightful, balanced, and deeply thoughtful in his editorial suggestions. The book simply would not have been a finished product without his invaluable guidance.

Stan Rosen of the National Space Society brought new life to the project just as I was placing it on the back burner to concentrate on other works. His unstinting support and generosity with his time prompted funding from people outside the publishing marketplace to underwrite the book's creation, resulting in an unusual but innovative commissioning process, and to him I will be forever grateful. From soliciting funds to technical input on the manuscript to marketing support, Stan is one of those rare individuals who does exactly what he says he will. He is a leader of the highest order.

Bruce Pittman, also of the National Space Society, was similarly helpful in the process, handling agreements and revisions despite heavy commitments for his work with NASA and the NSS. Bruce's leadership of the NSS made navigating sometimes murky waters almost easy, and I am grateful not just for the sake of *Space 2.0*, but for all the members of the NSS. My deepest thanks.

Mark Hopkins has been deeply and meaningfully engaged with the NSS and its precursor organization, the L5 Society, since the 1970s. He gave freely of his time for interviews and other input for the book, and his leadership of the NSS over many years is to be lauded.

Buzz Aldrin stepped up to write a wonderful foreword from a perspective that only he can claim. He is asked by many to provide much, and it is a testament to his belief in Space 2.0 that he took the time to support the book in a way that only he can. He is a brilliant man with ideas and program concepts that could carry us well into the twenty-second century. I am humbled by his participation in this project.

Jim Vaughan has been a friend for many years, and I have admired his illustration work for far longer than I have known him. His diligent and detail-oriented art speaks for itself, and I am grateful for its inclusion in this book. Jim infuses his space illustrations with a sense of optimism and a dynamic flair that is rare in this field, and he speaks to the viewer's heart as few can.

Edwin Sahakian is one of the book's key financial supporters and is credited as such, but he is also a newfound friend with a deep interest in the future of space exploration. Owner of a very successful air freight company, Edwin was among the first to step forward when we began the search for underwriting for this important project. His enthusiasm for aerospace and spaceflight is infectious and has doubtless inspired the many students he teaches in his second career as an academician.

To the many folks engaged in opening the space frontier as professionals who generously granted me time for interviews, my deepest thanks. It is not uncommon for such people, all leaders in their fields, to be hesitant to participate in book interviews—you never know how your statements will be reflected or interpreted after the fact. I appreciate the vote of confidence from these highly placed professionals. I've done my best to represent them as accurately and thoroughly as time and space allowed. I wish I could have included more of their statements, but books are limited with regard to word counts, so some of their thoughts, observations, and opinions are reflected more strongly in summary text than in direct quotes. My thanks to all of them.

Thanks are also due to the National Space Society, and its many thousands of members and hardworking volunteers, for giving people like myself a second home. While I have enjoyed my various stints with NASA and other spaceflight organizations, the NSS provides a forum to exchange thoughts and ideas that is incredibly beneficial. There are thousands of well-informed individuals in the organization, and many of them contributed to this book either directly or indirectly. To each, my thanks.

Finally, my deepest thanks to Aggie Kobrin, Mike Kobrin, Dave Dressler, Geoffrey Notkin, Howard Bloom, Lynne Zielinski, Art Dula, Buckner Hightower, Ronnie Lajoie, Joe Redfield, Fred Becker, Hoyt Davidson, Gary Barnhard, Dave Stuart, Larry Ahearn, Dale Skran, Al Anzaldua, Madhu Thangavelu, Ken Hoagland, Kirby Ikin, Carol Stoker, Greg Autry, Rhonda Stevenson, John Mankins, Naté Sushereba, Chantelle Baier, Ken Hoagland, and the many other volunteers who work hard each year to keep the dream alive via the International Space Development Conference. Without their untiring work, the NSS would simply not be what it is. The International Space Development Conference® (ISDC) is a place to meet, network, learn, and have a few much-needed laughs. Along with a fine quarterly magazine, *Ad Astra*, which I am privileged to edit, the ISDC is a primary benefit of belonging to the NSS. If you love space but have not been to the conference, you owe it to yourself to attend.

SPECIAL THANKS TO

Rod Pyle and the National Space Society would like to extend special thanks to the many supporters who made this book possible through their generosity.

Edwin Sahakian
John Saunders
Jim Haislip
Ben Bova
Bruce Pittman
Stan Rosen

The manuscript was reviewed and fact-checked by Stan Rosen, Bruce Pittman, and Mark Hopkins. Administration of the project was handled by Stan Rosen, Bruce Pittman, and Dean Larson. Susan Holden Martin provided invaluable insights, sourced many of the interviewees, and solicited additional funds. The administrative team also helped to arrange the many interviews and correspondences needed to make the book a success. Special thanks to Alla Malko for her assistance with Russian contacts and translations.

LIST OF INTERVIEWEES

Michael Alegria-Lopez
Chad Anderson
Alfred Anzaldua
Howard Bloom
Tory Bruno
Franklin Chang-Díaz
Tim Cichan
Leonard David
Jason Dunn
Pascale Ehrenfreund
Charles Elachi

Bill Gerstenmaier
Jeff Greason
Lori Garver
Mark Hopkins
Hiro Iwamoto
Steve Jurvetson
Jim Keravala
Pascal Lee
David Livingston
Robert Meyerson
Scott Pace

Bruce Pittman
Stan Rosen
Gwynne Shotwell
Dale Skran
Bill Tarver
Rick Tumlinson
Jakob van Zyl
George Whitesides
Robert Zubrin

NOTES

CHAPTER 1: SPACE 1.0, THE FINAL ACT

1   Open letter from Neil Armstrong, James Lovell, and Eugene Cernan to President Obama, published April 15, 2010. Retrieved from *Guardian* online on August 7, 2017. www .theguardian.com/commentisfree/cifamerica/2010/apr/15/obama-nasa-space-neil-armstrong.

2   O'Kane, Sean. "NASA buys two more seats to the International Space Station on Russia's Soyuz rocket." *The Verge*, February 28, 2017. www.theverge.com/2017/2/28/14761058/nasa-russia-spacex-soyuz-seats-contract-space-station-iss. Accessed 7/18/18. In addition to carrying up to seven astronauts, the shuttle was also capable of lofting over 50,000 pounds of cargo to orbit in a single flight. The Soyuz cargo module, called Progress—which does not carry any astronauts—can deliver about a tenth of that, or about 5,500 pounds.

3   Alabama, Texas, Utah, Louisiana, and Florida are the primary beneficiaries of spending on Orion/SLS.

4   This plan is in some flux as of late 2018, and SpaceX is developing capabilities far beyond what is needed for basic resupply of the ISS. But if both systems reach fruition, this is considered to be the logical path forward.

CHAPTER 2: A DARK AND FORBIDDING PLACE

5   There have been four deaths in the Russian program over the years, and two US space shuttles were lost carrying a total of fourteen astronauts.

6   This is generally true for human spaceflight. For many unmanned launches, the rockets can also lift off to the north or the south, entering what is called a polar orbit that follows a track across the North and South Poles as Earth turns below.

7   The effects of long-term radiation on the brain are still being explored. NASA's ongoing research is detailed here: www.nasa.gov/mission_pages/station/research/experiments/137. html. Retrieved August 11, 2017.

8   Recent research has suggested that this may be related to factors other than zero-g, possibly including elevated levels of carbon dioxide. In any case, it is another result of living off Earth.

CHAPTER 3: WHY SPACE?

9 Stirone, Shannon. "The Real Cost of NASA Missions." *Popular Science*, November 2015.

10 Dittmar, M. L. "Public Perception and NASA: Are We Asking the Right Questions?" Houston: Dittmar Associates, 2007.

11 Lefleur, Claude. "Cost of US Piloted Programs." *Space Review*, March 2010. The estimate might go up by $5–10 billion for 2018. In contrast, the International Space Station cost about $150 billion through 2015.

12 Wile, Rob. "Here's How Much it Would Cost to Travel to Mars." Quoting planetary scientist Pascal Lee. May 8, 2017.

13 Capaccio, Anthony. "F-35 Program Costs Jump to $406.5 Billion in Latest Estimate." *Bloomberg.com*, July 10, 2017. www.bloomberg.com/news/articles/2017-07-10/f-35-program -costs-jump-to-406-billion-in-new-pentagon-estimate. Retrieved August 11, 2017.

14 If you want more evidence, see the publication *NASA Spinoffs*, which has been in print (and is now online at spinoff.nasa.gov/Spinoff2008/tech_benefits.html) since the early days of the space program.

15 Interview with the author, October 2016.

16 Interview with the author, November 2016.

17 Rosen, Stanley. "Mind in Space." *USAF Medial Service Digest*, Jan–Feb 1976; White, Frank. *The Overview Effect—Space Exploration and Human Evolution*. Houghton-Mifflin, New York, 1987.

18 *People* magazine interview with Ed Mitchell, April 8, 1974.

19 Interview with the author, November 2016.

20 Interview with the author, March 2017.

21 Interview with the author, November 2016.

22 Wolf, Nicky. "SpaceX Founder Elon Musk Plans to Get Humans to Mars in Six Years." *The Guardian*, September 28, 2016. Reporting on remarks made by Musk during the 2016 International Astronautical Congress in Guadalajara, Mexico.

23 "NASA's Griffin: 'Humans Will Colonize the Solar System.'" *Washington Post*. September 25, 2005.

24 Ghost, Pallab. "Hawking Urges Moon Landing to 'Elevate Humanity.'" BBC News, June 20, 2017. Citing remarks made at the Starmus Festival in Trondheim, Norway, 2016.

25 Interview with the author, November 2016. As with most science-based topics, there is controversy about this point. There are those who feel that Earth has an almost limitless ability to adapt to most any human endeavor short of nuclear war, but they are in the minority. For a primer on some of these arguments, see the glossary section of the Intergovernmental Panel

on Climate Change, "Climate Change 2007: Working Group II: Impacts, Adaptation and Vulnerability." www.ipcc.ch/publications_and_data/ar4/wg2/en/annexessglossary-a-d.html. Retrieved April 1, 2017.

26   Interview at the 2016 Recode Convention. *The Verge*, June 1, 2016.

27   Britt, Robert Roy. "The Top 3 Reasons to Colonize Space." *Space.com*, October 18, 2001.

28   Interview with the author, October 2016.

CHAPTER 4: THE FIRST SPACE AGE

29   Marks, Paul. "One thing spacecraft have never achieved—until now." *BBC Future*, June 1, 2016.

30   Rhian, Jason. "Why Dream Chaser Didn't Win the Bid for Commercial Crew." *Spaceflight Insider*, October 14, 2014. www.spaceflightinsider.com/missions/commercial/lower-technical-maturity-schedule-uncertainty-cited-nasa-select-dream-chaser-commercial-crew-program/. Accessed April 12, 2018.

CHAPTER 5: DESTINATIONS

31   The five Earth-moon Lagrange points are called L1, L2, etc. L1 is between Earth and the moon, and L2 is on the far side of the moon. Both are in cislunar space but behave uniquely. L3 is on the far side of the earth from the moon. L4 and L5 lead and trail the moon along its orbit.

32   Strauss, Mark. "How Will We Get Off Mars?" *National Geographic*, October 2015.

33   All orbiting space habitats beyond LEO must be shielded from radiation.

34   For more detail on Lagrange points, see Howell, Elizabeth. "Lagrange Points: Parking Places in Space." *Space.com*, August 21, 2017. For more info on advocacy regarding L5, see information on the L5 Society (now merged with the National Space Society) here: www.nss.org.

CHAPTER 6: THE HUMAN FACTOR

35   Interview with the author, January 2017.

36   See Webrew, Benjamin B. "History of Military Psychology at the U.S. Naval Submarine Medical Research Laboratory." Naval Submarine Medical Research Laboratory, Groton, Connecticut, 1979.

37   Sutton, Jeffrey, ed. "During the Long Way to Mars: Effects of 520 Days of Confinement (Mars500) on the Assessment of Affective Stimuli and Stage Alteration in Mood and Plasma Hormone Levels." National Institutes of Health report, PLoS One 9, no. 4 (2014): e87087. PMCID: PMC3973648.

38   "Measurements of Energetic Particle Radiation in Transit to Mars on the Mars Science Laboratory." *Science* 340, no. 6136 (May 31, 2013): 1080–4.

39   By comparison, a person on Earth receives the equivalent of about 1/1000 sievert per year.

40   Cherry, Jonathan D., Bin Liu, Jeffrey L. Frost, Cynthia A. Lemere, Jacqueline P. Williams, John A. Olschowka, M. Kerry O'Banion, and Douglas L. Feinstein. "Galactic Cosmic Radiation Leads to Cognitive Impairment and Increased Aβ Plaque Accumulation in a Mouse Model of Alzheimer's Disease." University of IIllinois, 2012.

41   *Revolutionary Concepts of Radiation Shielding for Human Exploration of Space.* NASA publication, National Technical Information Service, NASA/TM—2005–213688, 2005.

42   The estimated costs of a single pound to launch vary depending on the source. In 2008, NASA estimated the cost of launching a pound of mass by the shuttle at about $10,000. Source: www.nasa.gov/centers/marshall/news/background/facts/astp.html. Accessed September 1, 2017.

43   Green, Marc, Justin Hess, Tom Lacroix, Jordan Marchetto, Erik McCaffrey, Erik Scougal, and Mayer Humi. "Near Earth Asteroids: The Celestial Chariots." Worcester Polytechnic Institute, Worchester, MA, 2013.

44   Rodjev, Kristina, and William Atwell. "Investigation of Lithium Metal Hydride Materials for Mitigation of Deep Space Radiation." Paper presented at the 46th International Conference on Environmental Systems, ICES-2016-124, July 2016, Vienna, Austria.

45   "Analysis of a Lunar Base Electrostatic Radiation Shield Concept." ASRC Aerospace Corporation study, MAC CP 04-0 1 Phase I Advanced Aeronautical/Space Concept Studies, 2004.

CHAPTER 7: SPACE ENTREPRENEURS

46   Bower, Tom. "Lost in Space." *Sunday Times*, January 26, 2016.

47   NTSB Press Release, July 28, 2015. www.ntsb.gov/news/press-releases/Pages/PR20150728.aspx.

48   Foust, Jeff. "How big is the market for small launch vehicles?" *Space News*, April 11, 2016.

49   Howell, Elizabeth. "How to Poop in Space: NASA Unveils Winners of the Space Poop Challenge." *Space.com*, February 15, 2017.

CHAPTER 8: SPACE EXPLORATION TECHNOLOGIES CORP.

50   Masunaga, Samantha. "SpaceX track record 'right in the ballpark' with 93% success rate." *Los Angeles Times*, September 1, 2016.

51   Sam Altman interview with Elon Musk for Y Combinator, September 2016.

52   Dillow, Clay. "The Great Rocket Race." *Fortune*, October 2016.

53   Brown, Alex. "Why Elon Musk Is Suing the U.S. Air Force." *The Atlantic*, June 5, 2014.

54   De Selding, Peter. "SpaceX's reusable Falcon 9: What are the real cost savings for customers?" *SpaceNews*, April 25, 2016.

55   This figure is quoted across a wide range. Lowest estimates fall at about $1,200 per pound via SpaceX, and go up to about $7,500 per pound for other launch providers. Part of this variation depends on other factors involved in the flight, such as military versus civilian oversight, type of cargo, and so forth.

56   Some launches of very heavy payloads require the Falcon 9 and Falcon Heavy rockets to be expendable, but SpaceX's upcoming BFR booster will be fully reused in all flights.

57   Chaikin, Andy. "Is SpaceX Changing the Rocket Equation?" *Air & Space Magazine*, January 2012.

58   Besides SpaceX, only Bezos's Blue Origin and, over twenty years ago, McDonnell Douglas's DC-X prototype have had substantive success in recovering reusable rockets. For more on DC-X, see Gannon, Megan. "20 Years Ago: Novel DC-X Reusable Rocket Launched into History." *Space.com*, August 16, 2003.

59   Wall, Michael. "Elon Musk Calls for Moon Base." *Space.com*, July 19, 2017. www.space.com/37549-elon-musk-moon-base-mars.html. Accessed June 10, 2017.

60   Etherington, Darrell. "SpaceX spent 'less than half' the cost of a new first stage on Falcon 9 relaunch." *Techcrunch*, April 5, 2017. techcrunch.com/2017/04/05/spacex-spent-less-than-half-the-cost-of-a-new-first-stage-on-falcon-9-relaunch. Accessed June 7, 2017.

61   Koren, Marina. "What Would Flying from New York to Shanghai in 39 Minutes Feel Like?" *The Atlantic*, October 2017.

62   Messier, Doug. "NASA Analysis: Falcon 9 Much Cheaper than Traditional Approach." *Parabolic Arc*, May 31, 2011.

63   Grush, Loren. "NASA is saving big bucks by partnering with commercial companies like SpaceX." *The Verge*, November 10, 2017.

64   Chaikin, Andy. "Is SpaceX Changing the Rocket Equation?" *Air & Space Magazine*, January 2012. www.airspacemag.com/space/is-spacex-changing-the-rocket-equation-132285884/.

CHAPTER 9: A NEW SPACE RACE

65   "S. Dade's Best and Brightest." *Miami Herald*, July 4, 1982.

66   Foust, Jeff. "Bezos Investment in Blue Origin Exceeds $500 Million." *Space News*, July 18, 2014.

67 Klotz, Irene. "Bezos is selling $1 billion of Amazon stock a year to fund rocket venture." *Reuters Business News*, April 5, 2017.

68 Fernholtz, Tim, and Christopher Groskopf. "If Jeff Bezos is spending a billion a year on his space venture, he just started." *Quartz*, April 12, 2017.

69 Foust, Jeff. "Blue Origin's New Shepard Vehicle Makes First Test Flight." *Space News*, April 30, 2015.

70 Clark, Stephan. "ULA chief says Blue Origin in driver's seat for Vulcan engine deal." *Spaceflight Now*, April 18, 2017.

71 Clark, Stephen. "Blue Origin details new rocket's capabilities, signs first orbital customer." *Spaceflight Now*, March 7, 2017. spaceflightnow.com/2017/03/07/blue-origin-details-new -rockets-capabilities-signs-first-orbital-customer. Accessed August 14, 2017.

72 As highlighted on Blue Origin's website: www.blueorigin.com/payloads. Accessed April 24, 2018. Also, see "NASA Funds Flight for Space Medical Technology on Blue Origin." NASA Feature News. www.nasa.gov/feature/nasa-funds-flight-for-space-medical-technology-on-blue -origin. Accessed April 24, 2018.

73 Interview with the author, September 2016.

74 Davenport, Christian. "An exclusive look at Jeff Bezos's plan to set up Amazon-like delivery for 'future human settlement' of the moon." *Washington Post*, March 2, 2017.

75 Ibid.

76 SpaceX launch costs are published on their website: www.spacex.com/about/capabilities. Accessed April 26, 2018. ULA does not publicly disclose launch fees.

77 Interview with the author, October 2016.

78 Ray, Justin. "ULA Gets Futuristic." *Spaceflight Now*, April 16, 2015.

CHAPTER 10: INVESTING IN SPACE

79 Sheetz, Michael. "Space companies received $3.9 billion in private investment during 'the year of commercial launch.'" *CNBC*, January 18, 2018. www.cnbc.com/2018/01/18/space -companies-got-3-point-9-billion-in-venture-capital-last-year-report.html.

80 "Space: The Next Investment Frontier." Goldman Sachs (podcast), May 22, 2017. www .goldmansachs.com/our-thinking/podcasts/episodes/05-22-2017-noah-poponak.html. Accessed July 26, 2018.

81 Folgar, Carlos, and Jess McCuan. "Private space exploration: the new final frontier." Quid, April 24, 2017. quid.com/feed/private-space-exploration-the-new-final-frontier.

82 "Space Investment Quarterly, Q4, 2017." Space Angels Holdings, January 18, 2018.

83 Masunaga, Samantha. "Saudi Arabia to invest $1 billion in Virgin Galactic." *Los Angeles Times*, October 26, 2017.

84 Interview with the author, September 2016.

85 Miller, Charles. "Affording a Return to the Moon by Leveraging Commercial Partnerships." Next Gen Space LLC, 2015.

86 Interview with the author, October 2016.

87 Foust, Jeff. "How Long Will the Money Keep Flowing?" *Space Review*, February 5, 2018. Quoting Shahin Farshchi, Lux Capital, in an address to the Space Tech Summit 2018 in San Mateo, CA.

CHAPTER 11: THE SPACE BETWEEN NATIONS

88 Sagdeev, Roald, and Susan Eisenhower. "United States–Soviet Space Cooperation During the Cold War." *NASA 50th Anniversary Magazine*, 2008. www.nasa.gov/50th/50th_magazine/coldWarCoOp.html. Accessed September 24, 2017.

89 Spudis, Paul. "JFK and the Moon." *Air & Space*, November 2013.

90 Seitzen, Frank. "Soviets Planned to Accept JFK's Joint Lunar Mission Offer." *Space Daily*, October 2, 1997.

91 Yam, Philip. "The Forgotten JFK Proposal." *Scientific American*, October 2012.

92 The full text of the OST can be found on the UN's Office for Outer Space Affairs website, www.unoosa.org/oosa/en/ourwork/spacelaw/treaties/introouterspacetreaty.

93 Committee on Aeronautical and Space Sciences. "Soviet Space Programs, 1971–75." Washington, DC: U.S. Government Printing Office, 1976.

94 There had been unmanned cooperation in the meantime. In 1977 and again in 1979, US experiments were carried aloft in Soviet Kosmos satellites.

95 Estimates of the US costs for the ISS often do not include the shuttle flights to build the station or the buildup of the shuttle program, which launched most of the heaviest components. The best estimate, which includes direct launch costs of the shuttle in support of the ISS, can be found here: "Final Memorandum, Audit of NASA's Management of International Space Station Operations and Maintenance Contracts." NASA Office of Inspector General, July 15, 2015. oig.nasa.gov/audits/reports/FY15/IG-15-021.pdf. Accessed September 25, 2017.

96 Pyle, Rod. "Path to Mars Should be Flexible, Experts Agree." *Space.com*, September 22, 2016.

97 For full text of ITAR, see: US Department of State, Directorate of Defense Trade Controls, International Trade in Arms Regulations. www.pmddtc.state.gov/regulations_laws/itar.html. Accessed September 24, 2017.

98 Ibid.

99   A specific list of ITAR-regulated nations and the level of proscription can be found on Stanford University's "Export Controlled or Embargoed Countries, Entities and Persons" list on the web: https://doresearch.stanford.edu/research-scholarship/export-controls/export-controlled-or-embargoed-countries-entities-and-persons#countries. Accessed May 27, 2018.

100   Interview with the author, November 2016.

101   "Cassini Quick Facts." NASA/JPL Cassini Legacy website. saturn.jpl.nasa.gov/mission/grand-finale/cassini-quick-facts/. Accessed September 19, 2017.

102   Vasani, Harsh. "How China Is Weaponizing Outer Space." *The Diplomat*, January 19, 2017.

103   Interview with the author, June 2017.

104   Interview with the author, December 2016.

105   Interview with the author, November 2016.

106   Stroikos, Dimitrios. "China, India in Space and the Orbit of International Society." White paper for the London School of Economics, 2016.

CHAPTER 12: THE RUSSIAN JUGGERNAUT

107   For more detail on the USSR's and Russia's long history in space, see Anatoly Zak's Russian Space Web, www.russianspaceweb.com.

108   The current cost to send US astronauts to the ISS aboard Soyuz is about $76 million and is estimated to rise to $80 million plus after 2018.

109   There have been numerous and well-documented launch failures of Russian rockets in the past decade, but most were due to faulty upper stages, not the primary booster stage.

110   "Russia's A5V moon mission rocket may be replaced with new super-heavy-lift vehicle." *RT* (formerly *Russia Today*), August 22, 2016.

111   Siddiqi, Asif. "Russia's Space Program Is Struggling Mightily." *Slate*, March 21, 2017.

112   Ibid.

113   Messier, Doug. "Russian Launch Failures Aren't a Bug, They're a Feature." *Parabolic Arc*, March 16, 2018.

CHAPTER 13: THE CHINA WILDCARD

114   Impey, Chris. "How China Entered the Space Race." *WIRED*, April 2015.

115   For more information on the Shuguang project, see *Encyclopedia Astronautica*, www.astronautix.com/s/shuguang1. Accessed September 10, 2017.

116   Nowakowski, Tomasz. "In the Footsteps of SpaceX: A Chinese company eyes the development of a reusable launch vehicle." *Astrowatch*, September 2017.

117 Stone, Christopher. "US cooperation with China in space: Some thoughts to consider for space advocates and policy makers." *Space Review*, February 2013.

118 "Full text of white paper on China's space activities in 2016." Translated from the original Chinese by the State Council of the People's Republic of China. english.gov.cn/archive/white_paper/2016/12/28/content_281475527159496.htm. Accessed April 28, 2018.

CHAPTER 14: TRUCK STOPS IN SPACE

119 Interview with the author, October 2016.

120 Landau, Damon F., James M. Longuski, and Buzz Aldrin. "Continuous Mars Habitation with a Limited Number of Cycler Vehicles." *Journal of the British Interplanetary Society*, April 2007.

121 See the full document at the NSS's website at: www.nss.org/legislative/positions/NSS-DFJ-Workshop-Recommendations-Nov-2016.pdf. Accessed October 19, 2017.

122 Vedda, James. "National Space Council: History and Potential." Center for Space Policy and Strategy, November 2016.

123 Garber, Steven. "The Space Exploration Initiative." NASA history archive. history.nasa.gov/sei.htm. Accessed April 28, 2018.

124 Interview with the author, January 2017.

125 Interview with the author, October 2017.

126 Foust, Jeff. "Moon Express wins U.S. government approval for lunar lander mission." *Space News*, August 3, 2016. spacenews.com/moon-express-wins-u-s-government-approval-for-lunar-lander-mission/. Accessed July 27, 2018.

127 Foust, Jeff. "National Space Council calls for human return to the moon." *SpaceNews*, October 5, 2017. spacenews.com/national-space-council-calls-for-human-return-to-the-moon. Accessed October 17, 2017. To see the directive, SPD-2, on the web: www.whitehouse.gov/briefings-statements/president-donald-j-trump-reforming-modernizing-american-commercial-space-policy/. Accessed June 1, 2018.

CHAPTER 15: DEFENDING EARTH

128 The Paleogene Period, which spanned 65 to 23 million years ago, was the kickoff for the Cenozoic Era, which continues into the present time.

129 "Asteroid of one-kilometer or larger strikes Earth every 600,000 years." *MIT Tech News*, September 10, 2003.

130 Lori Garver, interview with the author, 2016.

131 Interview with the author, October 2016.

132 For more information on planetary defense, see the NASA Planetary Defense FAQ, www.nasa .gov/planetarydefense/faq. Accessed October 21, 2017.

133 Van den Bergh, Sidney. "Life and Death in the Inner Solar System." Publications of the Astronomical Society of the Pacific, 101 (May 1989): 500–09. Updated by Nick Strobel, in "Effects of an Asteroid Impact on Earth." *Astronomy Notes*, 2013.

134 This program is referred to collectively as AIDA (Asteroid Impact and Deflection Assessment), and more information can be found at NASA's AIDA website: www.nasa.gov /planetarydefense/aida. Accessed October 30, 2017.

135 "Laser Bees." The Planetary Society website: www.planetary.org/explore/projects/laser-bees. Accessed October 20, 2017.

CHAPTER 16: SETTLING THE FINAL FRONTIER

136 Spudis, Paul. "Why We Go to the Moon." *Air & Space/Smithsonian*, October 17, 2017.

137 Shukman, David. "Humans at risk of lethal 'own goal.'" BBC News, January 19, 2016.

138 Interview with the author, June 2017.

139 For a full transcript of the speech, see: Mosher, Dave. "Here's Elon Musk's complete, sweeping vision on colonizing Mars to save humanity." *Business Insider*, September 29, 2016.

140 Ibid.

141 Davenport, Christian. "Jeff Bezos on nuclear reactors in space, the lack of bacon on Mars and humanity's destiny in the solar system." *Washington Post*, September 15, 2016.

142 Interview with the author, November 2016.

143 "NASA's Griffin: 'Humans Will Colonize the Solar System.'" *Washington Post*, September 25, 2005.

144 "Summary of H.R. 4752—Space Exploration, Development and Settlement Act of 2016." www.congress.gov/bill/114th-congress/house-bill/4752. Retrieved September 20, 2017.

145 O'Neill, Gerard K. *The High Frontier: Human Colonies in Space*. Space Studies Institute, 1977, 2013.

146 O'Neill, Gerard K. "The Colonization of Space." *Physics Today* 27 (9): 32–40.

147 John, R. D., and C. Holbrow, eds. "Space Settlements: A Design Study." NASA, SP-413, Scientific and Technical. NASA Ames Research Center library, 1975.

148 Dyson, Freeman J. "Shells Around Suns May Have Been Built." *Science News Letter*, June 18, 1960, page 389.

149 Dickens, Peter. "The Humanization of Space: To What End?" *Monthly Review* 62, no. 6 (November 2010).

150 "Deep Space Gateway to Open Opportunities for Distant Destinations." NASA News, March 2017. www.nasa.gov/feature/deep-space-gateway-to-open-opportunities-for-distant -destinations. Accessed April 28, 2018.

151 Norris, Guy. "NASA Takes First Deep Space Propulsion Steps." *Aviation Week*, July 19, 2017. aviationweek.com/space/nasa-takes-first-deep-space-propulsion-steps. Accessed September 30, 2017.

152 "Moon Village: Humans and Robots Together on the Moon." European Space Agency, DG's News and Views, March 1, 2016. www.esa.int/About_Us/DG_s_news_and_views/Moon _Village_humans_and_robots_together_on_the_Moon. Accessed September 30, 2017.

153 Ibid.

154 Geggel, Laura. "City-Size Lunar Lava Tube Could House Future Astronaut Residents." *Live-Science*, October 2017.

155 "Marius Hills pit offers potential location for lunar base." NASA Solar System Exploration Research Virtual Institute. sservi.nasa.gov/articles/lava-tube-lunar-base. Accessed October 20, 2017.

156 2017 International Astronautical Conference in Adelaide, Australia. Details at the SpaceX website: www.spacex.com/mars. Accessed May 1, 2018.

157 The National Space Society has its own, more detailed Space Settlement Roadmap online at space.nss.org/space-settlement-roadmap/. Accessed May 13, 2018.

158 Jim Keravala address to the first Space Settlement Summit in Los Angeles, California, October 2016.

159 Interview with the author, August 2016.

160 Address to the National Space Society's Space Settlement Summit, October 2017.

161 Interview with the author, November 2016.

162 Kennedy, Brian. "Five facts About Americans' views on space exploration." Pew Research Center, July 2015.

163 Media outlets from the *Washington Post* (owned by Jeff Bezos), to the *National Review*, to *Al Jazeera*, reported positively on the Falcon Heavy launch and the reactions of people worldwide.

CHAPTER 17: A NEW AGE DAWNS

164 Address by Elon Musk at NASA's "ISS R&D conference" in Washington, DC, July 19, 2017.

165 "ABC News Special Report." Interview of Elon Musk by David Kerley from a press conference after launch of the Falcon Heavy, February 6, 2018.

CHAPTER 18: YOUR PLACE IN SPACE 2.0

166 Interview with the author, October 2016.

GLOSSARY

Apollo program: The US space program started in the 1960s to allow American astronauts to land on, explore, and return from the moon. The Apollo program ended in 1975 with the Apollo-Soyuz Test Program.

Atlas V, Delta IV, and Vulcan: United Launch Alliance's (ULA's) family of rockets that provide commercial and government launch services. The Vulcan is currently being designed; the Atlas and Delta vehicle families have been flying for years.

Blue Origin: A spaceflight company founded by entrepreneur Jeff Bezos in 2000 to build reusable rockets and spacecraft for launch services and suborbital tourist flights.

Clean-sheet design: A complete redesign of a project or piece of hardware starting with a "clean sheet" of paper and new thinking about the requirements to be met.

CNSA: The Chinese National Space Agency.

COTS: Commercial Orbital Transportation Services, a program implemented by NASA in 2006 to demonstrate the efficacy of NASA working with commercial partners in then nontraditional relationships.

CRS: A NASA program called Commercial Resupply Services that ran from 2008 to 2016, which granted a series of contracts to deliver cargo to the International Space Station on private rockets and spacecraft. The first CRS contract was awarded to SpaceX in 2016.

ESA: The European Space Agency, an organization of European nations that undertakes their collective spaceflight ventures.

Falcon 9 and Falcon Heavy: Rockets built by SpaceX that provide commercial and government launch services.

Gateway: New name for the Lunar Orbiting Platform-Gateway (see entry below).

289

GCR: Galactic Cosmic Radiation, also known as cosmic rays. This is radiation from beyond the solar system that has a negative impact on both human bodies and electronics in space.

ISRU: In-Situ Resource Utilization, the use of resources found in space to support operations there. Refers primarily to atmospheric gases, water ice, and metals found on the moon, asteroids, and planets such as Mars that can be refined into useful substances.

ISS: International Space Station, built by NASA and a consortium of multinational partners operating under NASA leadership. The ISS began crewed operations in 2000.

ITAR: International Traffic in Arms Regulations, a set of US government regulations to control the sale of sensitive technologies to foreign countries. First enacted in 1976.

JAXA: The Japanese national space agency.

JPL: The Jet Propulsion Laboratory, the NASA field center responsible for the robotic exploration of space.

Low Earth Orbit or LEO: The orbital region above Earth that extends roughly from the lowest possible stable orbit to about 1,200 miles in altitude. This is a region in which the International Space Station orbits (at about 250 miles) and the space shuttle operated (190 to 330 miles). It is also quite heavily populated with satellites.

Lunar Orbiting Platform-Gateway: Also known as LOP-G, or more simply the Gateway, it is a planned station orbiting the moon that will host periodic crews from Earth. Intended to be built via a coalition led by NASA in possible cooperation with Russia and other nations.

NASA: The National Aeronautics and Space Administration, the US's primary space exploration organization.

National Space Council: A US governmental body that reports directly to the president on matters of space policy and operations. First founded under the Eisenhower administration in the 1950s, the NSC is now in its third iteration, having been reinstated by the Trump administration.

NEO: Near Earth Object, referring to asteroids, comets, and other bits of space debris that have orbits coming close to Earth.

New Shepard, New Glenn: Rockets being built and tested by Blue Origin to provide commercial and government launch services. The New Shepard is specifically designed to provide tourist flights to suborbital altitudes.

NewSpace: Refers broadly to a new generation of companies making private investment in space activities across the board. One suggested downside of the term is that it draws an unspoken comparison to "OldSpace," the traditional large aerospace companies, which some find objectionable.

NSS: The National Space Society, a grassroots space advocacy group dedicated to the long-term settlement of space and the use of space resources for the betterment of humanity.

Orion: A capsule intended primarily for crewed missions beyond low Earth orbit.

Payload: The cargo lofted into space by a rocket. Can refer to people or hardware.

Pressurized: Characteristic of a part of a spacecraft that contains a gas, generally a breathable atmosphere, but can also refer to any gas used to fill a vacuum.

Private spaceflight: A broad definition would include any private company engaged in spaceflight support or services. A subset of this would include space entrepreneurs—those who have invested their own money into spaceflight.

Propellant: Substances that when accelerated away from a rocket create thrust.

Reusability: In this context, referring to rockets and spacecraft that can be flown, refurbished, and reflown to save money.

Roscosmos: A state corporation responsible for the spaceflight and cosmonautics program for the Russian Federation. Roscosmos runs the majority of Russian (and previously Soviet) space programs.

Shenzhou and Tiangong: Respectively, the collective names for China's new space capsules and space stations to provide crewed spaceflight capability.

SLS: Space Launch System, a large rocket being built by NASA primarily for missions beyond low Earth orbit. It is derived from technology designed for the space shuttle and Constellation programs, the latter of which were intended to return US astronauts to the moon. Constellation ran from 2004 to 2010.

Soyuz: A Russian spacecraft family (both capsule and rocket share the name, but the term most commonly refers to the capsule) designed in the 1960s to take cosmonauts to Earth orbit and the moon.

Variants of the Soyuz continue to fly today, primarily in support of the ISS.

Space: The unofficial definition of "space" is the region above the Karman line, anything beyond sixty-two miles (100 km) in altitude.

Space 2.0: The new era of spaceflight that started between 2000 and 2002. This span includes the first crew sent to the International Space Station in 2000, but also the founding of Blue Origin, also in 2000, and SpaceX, in 2002. Generally refers to the entry of a new generation of private entrepreneurs into large spaceflight activities, such as the building of rocket boosters, and more broadly to the legions of individuals and companies now becoming involved in the space sector. While not excluding NASA and other space agencies, Space 2.0 is defined by an increasing reliance by NASA on private companies in new and unique partnerships that place increased risk on the private entities. Not completely interchangeable with NewSpace, which is defined as companies having begun in the early 1980s, after the Apollo program, but is often used in the same context.

Space entrepreneurs: A person or group of people who have invested their own money into the development of spaceflight hardware, services, or other businesses. These can include rockets, spacecraft, robots, software, spacecraft components, ground support, and more. In general use, the term applies most commonly to businesspeople such as Elon Musk, Jeff Bezos, Paul Allen, Richard Branson, and others who have created companies to build rockets, spacecraft, or space technology for commercial applications.

Space infrastructure: A term broadly describing the hardware and organization needed to operate routinely and efficiently in space. The first elements are already in place—rocket launch facilities, control centers, and so forth. The evolution of this concept includes, for example, crewed orbital outposts, refueling depots, and off-Earth facilities for mining, extracting, refining, and storing supplies needed for extended operations in space.

Space race: Competition between nations for achievement in space. Specifically, a period during the 1960s when the US and Soviet Union were both building space programs to land humans on the moon and return them to Earth. Can also refer to the "new space race," the period now upon us, in which private companies are competing with each other, and with foreign entities, to provide increasingly affordable launch and spaceflight services. Some experts suggest another space race coming between national programs such as those of China and India.

Space settlement: The movement of human civilization into Earth orbit and beyond. Space settlement strives for permanent human presence in space and the use of space to improve life on Earth.

Space shuttle: Also known as the Space Transportation System, or STS, the shuttle was the US's primary system for flying humans and cargo into space from 1981 through 2011.

SpaceX: Corporate identity of Elon Musk's Space Exploration Technologies, which builds and flies rockets and spacecraft. Founded in 2002.

ULA: United Launch Alliance, a joint venture via a partnership of Lockheed Martin and the Boeing Company, founded in 2006. To date, ULA has flown uncrewed spacecraft aboard its expendable rockets—the Atlas V and the Delta IV.

Zero-g, weightlessness, and microgravity: Generally interchangeable terms referring to floating in space; the condition in which astronauts and space hardware do not significantly experience the effects of gravity.

INDEX

Note: Page numbers in *italic* refer to photos or illustrations and accompanying captions

A

AAS (American Astronautical Society), 267

Abbey, George, 195

ACES. *see* Advanced Cryogenic Evolved Stage

acid rain, 228

active radiation shielding, 90–91

Adams, Douglas, 261

Ad Astra magazine, 266

Aditya satellite, 177

Advanced Cryogenic Evolved Stage (ACES), *145,* 145–148, 208–209

Advanced Development Programs, 138

Advanced Programs (ULA), 147

Aerospace Corporation, 125

AIAA (American Institute for Aeronautics and Astronautics), 267

Airbus A380 jumbo jet, 129

air circulation, 74, *74*

airliners, commercial, 120, 245

Aldrin, Buzz, *45,* 73, 168, 170, 211–212, *251*

Aldrin Cyclers, 211–212, *213*

Allen, Paul, 102, 151

Altair, *8*

aluminum, spacecraft construction material, 87, *87*

Alzheimer's disease, 86

Amazon.com, 35, 136, *136, 236*

American Astronautical Society (AAS), 267

American Institute for Aeronautics and Astronautics (AIAA), 267

Ames Research Center, 199

Anderson, Chad, *155,* 155–158, 220

Angara rocket, 186

Antarctic Treaty (1959), 165, 218

Antares rocket, 153, *153*

anti-satellite weapons (ASAT), *170*

Apollo program (in general)

capsules, 18, 19

cost of, 27

defined, 289

docking with Soyuz, *166*

and Gemini flights, 44

inspiration from, 31, 32

lunar landing program, xiii–xiv, 13, 44, 55, 65, 141, 214

Lunar Module, 141

radiation exposure, 82–83

reentering of, into Earth's atmosphere, *19*

replacement of, 13

scientific samples from, 219

Apollo 8 mission, *32,* 184

Apollo 11 mission, *45,* 168, 184

Apollo 13 mission, 7

Apollo 14 mission, 32

Apollo 16 mission, *71*

Apollo 17 mission, xiii, *23, 205*

Apollo-Soyuz Test Project (ASTP), 165–166

Apophis (asteroid), 230

Apple, Inc., 157

ARM (Asteroid Redirect Mission), *9,* 10

Armageddon (movie), 230

Arms Export Control Act, 169

Armstrong, Neil, xiii, 6, 7, *8, 45,* 73–74

artificial gravity, *80,* 80–82

ASAT (anti-satellite) weapons, *170*

Asteroid Redirect Mission (ARM), *9,* 10

asteroids, *64, 158*

    Chicxulub impactor, 223–224

    Chinese intentions of exploring, 197

    collecting resources from, *205,* 206, 218

    defending Earth from, 223–231, 248

    destruction of, *228–229,* 230, *231*

    detection of, *224*

    Hayabusa, *175*

    impact of, *227*

    mining resources from, *88,* 89

    as outposts, 64

    radiation exposure on, 83

    traveling to, 60

ASTP (Apollo-Soyuz Test Project), 165–166

astronauts, 69–91

    and artificial gravity, 80–82

    effects of cramped quarters on, 76–80

    and global perspective, 32

    medical experiments involving, 72–76

    and radiation, 82–91

    suits for, 69–71, *199*

    "taikonauts," 192, 193

ATK, 152

Atlantis, 2–7, 12

Atlas rocket, 11, 118, 122, *142, 164,* 175, 187

Atlas V, 289

    design of, 142

    preparing for launch, *143, 147*

    Russian parts on, 140, *149*

    sales of, 188

    and Starliner, 50

    US Air Force payload with, *119*

    as US rocket, 55

    Vulcan as replacement for, 135, 143

Atomic Energy Commission, 214

AT&T, 211

ATV (Automated Transfer Vehicle), 172

aurora, *61*

Ausbun, Kaylee, 115

Australia, 270

Australian Space Research Institute, 270

Automated Transfer Vehicle (ATV), 172

Axiom Space, 55

B

B-52 bomber, 96, *96*

Bangladesh, 180

BE-3 engine, 139

BE-4 engine, 139–140, *140*

BEAM (Bigelow Expandable Activity Module), 54, *54*

Bell Laboratories, 211

Bezos, Jeff
and Blue Origin, 135–136, *136,* 140–141, 236
future visions for Earth, 35–36
and international space agencies, 181
as space investor, 151, 154
and Virgin Galactic, 101–102

Bigelow, Robert, 54

Bigelow Aerospace, 246

Bigelow Expandable Activity Module (BEAM), 54, *54*

Bigelow expandable modules, 247

Bigelow habitation module, *237*

Bigelow test module, 55

Big Falcon Rocket, *86, 125,* 129, 244, 245, 249

Big Falcon Spaceship, 129, 244, 245, 249

"billionaires' club," 151, 154

Block 1 shuttle, *53*

Block 5 rocket, 126

Bloom, Howard, 34

Blue Moon lander, 141, *141,* 211, 218, 244, 246, 247

Blue Origin, 135–141
Advanced Development Programs, 138
BE-3 engine, 139
BE-4 engine, *138,* 139–140, *140*

Jeff Bezos and, 35–36, *136,* 236
Blue Moon lander, 211, 218, 247
cislunar infrastructure with, 246
defined, 289
Rob Meyerson and, *139*
New Glenn rocket, *136,* 244, 245
New Shepard, 51, *51, 137*
and new space race, 135–141
pilot space settlements by, 248
public support of, 252
and reusability, 186, 197, 204
as rival to Virgin Galactic, 101–102
rockets of, 55
suborbital test flights from, 127
for tourist flights, 59, 245
and United Launch Alliance, 187–188

Boeing
commercial space provider, 2, 158
late deadlines of, 7, 187
NASA partnerships with, 140, 141, 210
as private launch company, 9–11, *157,* 185
Starliner program, 12, 50, *50*

Bolden, Charles, *116*

bone-density loss, *72, 73*

Bowie, David, 260

Branson, Richard, 51, 94, 97, *97,* 100

breakthrough technologies, 217

Brezhnev, Leonid, 165–166

British Interplanetary Society, 268

Brookhaven National Lab, 86

Bruno, Tory, *143,* 143–145, 147, 148

Buran space shuttle, 184

Bureau of Political-Military Affairs, 169
Bush, George W., 6, 214

C

California Central Coast, *112*
caloric intake, 76
Caltech, 36
Canadarm robots, 179, *179*
Canadian Space Agency, 48, 113, 167
    accomplishments of, 179
      Canadarm, *179*
      NASA partnership with, 179
Canadian Space Society, 269
Cantrell, Jim, 104
carbon dioxide "bubbles," 74, *74*
Cardon, Thatcher, 108
Carina (constellation), *83*
Cassini-Huygens mission, 169–170
Cassini mission, 169–170, *170*
CASSIOPE satellite, 113, *133*
Centre national d'études spatiales (CNES),
   172
centrifugal force, 22, 80–81
centrifuges, *81,* 81–82
Cernan, Gene, 7, *8, 205*
Challenger disaster, 5–6
Chandrayaan satellites, 177
Chang-Díaz, Franklin, 32–33, *33,* 252
Chiao, Leroy, 195
Chicxulub impact, 223–224, *225*
Chinese National Space Agency (CNSA),
   191–200, 289
    budget of, 196
    crewed stations in LEO, 61, 62
    first trip to space, 48

    habitation modules created by, 247
    launch costs lowered by, 148, 246
    and Mars-500 study, 79
    and militarization of space, 170–171,
      220–221
    moon landing planned by, 65
    as new space power, xiv, 12, 188
    radiation guidelines of, 86
    and space entrepreneurs, 181
    stated intentions of, 196, 197
    Tiangong stations launched by, 54
    unsuccessful Mars voyages of, 177
Chryse Planitia, *250*
Chrysler corporation, 125
CisLunar-1000 Economy, *207,* 208,
   214–215
cislunar space, 60, *60,* 246–248
clean-sheet design, 289
Clinton, Bill, 167
CNES *(Centre national d'études spatiales),*
   172
CNSA. *see* Chinese National Space
   Agency
COLBERT space treadmills, *82*
Cold War, 12–13, 28, 42, 72, 141, 171,
   192, 198
Collins, Mike, xiii, 73
Colorado School of Mines, 147, 208
Columbia, 5
Columbus module, 172–173
comets, 224
commercial airliners, 120, 245
Commercial Crew Program, 135
commercialization of space, 212–217,
   234

Commercial Orbital Transportation
  Services (COTS), 130, 194, 215, 289
Commercial Resupply Services (CRS),
  215–216, 289
Commercial Space Launch
  Competitiveness Act (2015), 219
Conrad, Pete, *72*
Constellation program, 6–9, *8,* 141
Convair, *142*
cooperation, international. *see*
  international cooperation
COTS. *see* Commercial Orbital
  Transportation Services
cramped quarters, psychological effects
  of, 76–80
craters, 224, 227
Cretaceous Period, 223
crewed landings, 247
crewed stations, 61–62, 64, 184
CRS (Commercial Resupply Services),
  215–216, 289
cryogenic state (fuel), 148
cubesats, 103, *104–106,* 106, 160
Cultural Revolution (1966), 192
Cupola, *77*
Curiosity Mars rover, 84, *85*
Cygnus, *12,* 152, 153

D
DART mission (Double Asteroid
  Redirection Test), *228,* 228–229
deep-space economy, 206, 208–211
Deep Space Gateway, 241–242
Deep Space Industries, 180, 217, 218
defense systems, 248

Defense Trade Controls, 169
deflector shields, 90
Deimos, 60, 175
Delta IV Heavy, 142, *146*
Delta IV rocket, 52, 55, 135, 142, *146,*
  289
Delta rockets, 118, 122
delta-v, 19–20
Denmark, *156*
depression, in astronauts, 79
*Deutsches Zentrum für Luft- und
  Raumfahrt e. V.* (DLR), 172, 173
DFJ (Draper Fisher Jurvetson Venture
  Capital), 155, 158
Didymoon (asteroid), 230
Dittmar Associates, 26–27
DLR (*Deutsches Zentrum für Luft- und
  Raumfahrt e. V.*), 172, 173
dominance, 76–77
Double Asteroid Redirection Test
  mission (DART), *228,* 228–229
Dove satellite, *160*
Dragon 2, *11,* 19, 50, *50,* 55, 123, 130,
  159, 210
Dragon rocket, 119, *123,* 126, 130
Draper Fisher Jurvetson Venture Capital
  (DFJ), 155, 158
Dream Chaser, *11,* 52, *52,* 55
Dyson, Freeman, 36, 241
Dyson Sphere, 241

E
Earth orbit, 245–246
Earth suborbital (as space destination),
  59, 245–246

East India Trading Company, 125–126

Edwards Air Force Base, 95

Ehrenfreund, Pascale, 173–174

Elachi, Charles, 169–170

electrical fields, 90

Electron rocket, 103–104

elimination of waste, 70, 107

Endeavor, 3

Energia booster, 128

entrepreneurs. *see* space entrepreneurs

Epsilon rocket, 175

ESA. *see* European Space Agency

European Space Agency (ESA)
 accomplishments of, 172–174
 and crewed landings, 247
 defined, 289
 HTV cargo, *173*
 and international space agencies, 181
 lowering of launch costs by, 148
 modules, *166*
 Moon Villages, 65, 242
 and NASA, *170*
 partnerships with, 48, 167, 195
 successful trips to Mars, 177
 work on Huygens probe by, 169–170

EVAs. *see* Extra Vehicular Activities

Eve airplane, *95,* 96, 101

exercise, in space, 82

expeditionary model, settlement vs., 237

Explore Mars, 270

Explorer 1 satellite, 40, *40–41*

extended confinement, 76, *76*

Extra Vehicular Activities (EVAs), 70, *71,* 107

F

FAA. *see* Federal Aviation Administration

fairings, 119, 126, 194, *259, 261. see also* payloads

FAITH (Final Assembly, Integration, and Test Hanger), 95, 100

Falcon 1 (shuttle), 115, 126, 156, *156,* 158, *159,* 160

Falcon 9 (shuttle)
 for commercial launches, 194
 under construction, *215*
 cost of, 128, 130
 and COTS program, 215
 defined, 289
 first flight of, 126
 first launch of, 156
 with hypersonic grid fins, *127*
 Merlin engine for a, *131*
 and reusability, 39, 119, 259
 and SpaceX, 11, 112–115, *113–115, 118, 120–121*
 as US rocket, 55
 at Vandenberg A.F.B., *133*

Falcon Heavy
 construction of, 123
 defined, 289
 Delta IV Heavy and, 142, *146*
 first launch of, *129,* 255–257, *256, 258–261,* 261
 for interplanetary travel, 128
 New Glenn and, *136,* 137, 245
 reusability of, 259
 and SpaceX, 11
 as US rocket, 55

Farshchi, Shahin, 161

FCC (Federal Communications Commission), 131
feathering system, 98–99
Federal Aviation Administration (FAA), 99, 131, 220
Federal Communications Commission (FCC), 131
femtosats, 103
Final Assembly, Integration, and Test Hanger (FAITH), 95, 100
Firefly Aerospace, 103
flight surgeons, 73, 79, 108
Forbes magazine, 151
Founders Fund, 126
France, 79, 172–173
Freedom space station, 166, *166*
French postal services, 211
Friedman, Louis, 265
Friendship 7 spacecraft, *164*
fuel
    for achieving orbit, 16
    cost of, 128
    for leaving Martian surface, 63
    liquid methane as, 138
    for radiation shielding, 88
    for SpaceX rockets, 144
    storage of, 148
fuel depots, 159, 204, 208, 235
fuselages, 122, *123*

G

Gagarin, Yuri, 42, *42*
galactic cosmic rays (GCRs), 83, 90, 290
Garver, Lori, 30, *30,* 209, 226, 251
Gateway, 241–242, 246, 249, 289

GCRs (galactic cosmic rays), 83, 90, 290
Gemini 4 mission, *71*
Gemini spacecrafts, 44, *45, 71,* 192
genetic damage, from radiation, 86
George Washington University, 186–187
geostationary Earth orbit, 59–60, 130
geosynchronous Earth orbit, 59–60, 206
German Aerospace Center, 172–173, 176
Gerstenmaier, Bill, 168, 210–211, 242
getting involved, 263–271
    citizen groups, 264–265
    expanded opportunities for, 263–266
    indirect involvement, 264
    and the internet, 264
    organizations for, 266–271
Glenn, John, 42, 142, *164*
Goddard Space Flight Center, *31*
GOES weather satellite, *235*
Google, 160
government spaceflight, 246
GPS, 28
gravity
    artificial, *80,* 80–82
    on Earth, 20, *20*
    effects of, on astronauts, 22
    micro-, 82, 292
    overcoming, 16–18
Greason, Jeff, 217, 219, 251–252
Great Britain, 177, 211
Griffin, Mark, 35, 237, 238
Gross Space Product, *207*
Guardian, 98

H

H3 launch vehicle, 175
Hadfield, Chris, 179
"handshake in space," *166*
Harvard University, 36
Haughton-Mars Project Research
    Station, 270–271
Hawking, Stephen, 35, 235
Hayabusa probe, 175, *175*
Hayn Crater, *62*
heat-absorbing tiles, 18–19
high-altitude aircrafts, 69–70, *70*
*The High Frontier: Human Colonies in
    Space* (O'Neill), 240, 241
Highway 58 (California), 94
HI-SEAS simulator, *78*
The Hitchhiker's Guide to the Galaxy
    (Adams), 261
Hopkins, Mark, 36, 265
Hoshide, Akihiko, *28*
HTV spacecraft, *173–174*, 175
Hubble Space Telescope, 2
Hudson Bay Company, 251
Human Exploration and Operations,
    168, 210
humans
    destinations for, in space, 59–66
    and need for gravity, 81
    spacecrafts for transporting, 49–55
    in spaceflight, 44–48, 158, 177–178,
        264
    survival of, 20–21, 34–36, 234–236,
        238
human waste, elimination of, 70, 107
Huygens probe, 169–170

hypersonic grid fins, 127, *127*

I

IBM, 157
Icarus Interstellar, 270
ICBMs, 230
IKAROS spacecraft, 175
incentives, for space settlement, 250–252
Indian Space Research Organization
    (ISRO), 148
    accomplishments of, 176–178
    collaboration with Russia, 177–178
    lowering launch costs, 246
    Mangalyaan Mars orbiter, *177*
    Mars Orbiter Mission (MOM), *177*
    NASA partnership with, 177, 178
    PSLV rocket, *176*
    space entrepreneurs in, 181
Indian Space Society, 269–270
infrastructure. *see* space infrastructure
In-Situ Resource Utilization (ISRU), 290
Institute of Biomedical Problems, 79
International Astronautical Congress,
    236, 244
international cooperation, 163–181, *164,*
    266–271
    and international accomplishments,
        172–179
    and international space agencies,
        180–181
    ITAR, 168–170, 172
    space race as example of, 163–168
International Space Development
    Conference, 35

International Space Station (ISS), 53, *168*

  air circulation in, *74*
  aluminum hulls in, *87*
  Antares rocket and, *153*
  Apollo mission and, xiii–xiv
  Atlantis mission and, 2–3, 5–6
  aurora as seen from, *61*
  BEAM module on, *54*
  cargo on, *210*
  centrifuges on, *81*
  cost of, 168
  decommission and privatization of, 246
  defined, 290
  Dragon and, *50*
  Dream Chaser and, *52*
  and European Space Agency, 172
  as example of international cooperation, *48,* 163, 165, 195
  as Freedom station, 166, 167
  hardware on, 153
  HTV cargo, *173*
  Kelly brothers on, *75*
  André Kuipers on, *82*
  launching to, 15–16
  in low Earth orbit, 59
  Elon Musk and, 117
  NASA and, 65
  National Laboratory, 216
  permanent occupation of, *53*
  private space companies and, 9, 210
  shuttle approaching, *46*
  Soyuz and, *49, 185, 189*
  Starliner and, *50*
  3-D printers on, 107
  Unity module, *171*
  window structures aboard, 77, *77*
International Space University, 268
International Traffic in Arms Regulation (ITAR), 168–170, 172, 195, 198–199, 290
internet, 29, 130–132, 264
investors. *see* space investors
involvement in space, 263–271
Iran, 180
isogrids, 122, *123*
ISRO. *see* Indian Space Research Organization
ISRU (In-Situ Resource Utilization), 290
ISS. *see* International Space Station
ITAR. *see* International Traffic in Arms Regulation
Iwamoto, Hiro, 174–175

J

Jackson, Michael, 34
James Webb Space Telescope, *31*
Japanese National Space Agency (JAXA), 290
  accomplishments of, 174–175
  Hayabusa probe, *175*
  HTV cargo, *174*
  Mars missions of, 175, 177
  modules, *166*
  partnerships with, 48, 167, 175
  probing by, 243–244
Jet Propulsion Laboratory (JPL), 34, *41,* 169, 198, 239, 265, 290

Johnson Space Center, 30–31, 73–74, 195
JPL. *see* Jet Propulsion Laboratory
Jupiter, 20, 177, 197
Jurvetson, Steve, 126, 131–132, 155, 158–160, *159,* 212

K
Kelly, Mark, *75,* 75–76
Kelly, Scott, *73, 75,* 75–76
Kennedy, John F., 13, 42, *164,* 164–165
Kennedy Space Center, 6, 12, 30–31, 113, 255
Keravala, Jim, 250–251
Khrushchev, Nikita, *164,* 164–165
kinetic impactor, 229
Kistler Aerospace, 137
Knight, Peter, *70*
Kuipers, André, *82*
Kymeta, 107

L
L5 Society, 265
Lagrange points, 60, 64, 229, 237, 240, 248
Laser Bees, 230
LauncherOne, 100, 102
launch vehicles, 186, 187
lava tubes, *89, 243*
LEO. *see* low Earth orbit
Leonov, Alexei, *45, 166*
Link Space, 194
LMO (low Mars orbit), 60
Lockheed Martin, 50, *63,* 141, *142,* 144
Logsdon, John, 186–187

long-duration spaceflight, *75*
Long March boosters, 49, 192–193
LOP-G. *see* Lunar Orbiting Platform-Gateway
Los Angeles International Airport, 121
Lovell, Jim, 7
low bone density, *72*
low Earth orbit (LEO)
    aluminum hulls for shuttles in, *87*
    Apollo 17 in, *205*
    aurora in, *61*
    capsule design for, 55
    defined, 290
    exploration beyond, 10
    outposts in, 61–62
    and radiation, 21, 82–85
    satellites in, 107
    as travel destination, xiv, 7, 59
low Mars orbit (LMO), 60
Luna 2 spacecraft, 183
Lunar and Planetary Institute, 233
Lunar Module, 141
Lunar Orbiting Platform-Gateway (LOP-G), 187, *241,* 241–242, 247, 290
lunar rovers, *248*
Lunar Roving Vehicle, *205*
lunar space stations, 246–247
Luxembourg, 161, 180, 219
Luxembourg Space Cluster, 180

M
Made In Space, Inc., 106–107
Maezawa, Yusaku, 129
magnetic field, on Earth, 22, 83, 84, 90

Manber, Jeffrey, 106

Mangalyaan Mars orbiter, 177, *177*, 178

Mao Zedong, 192

Marius Hills, *243*, 243–244

Mars

    Aldrin Cyclers on, *213*

    as candidate for human spaceflights, 62–63, *63*

    capsules orbiting near, 55

    Chinese intentions of reaching, 197

    collecting resources on, 218

    colonies on, 34–35, *89*, *238*, 244

    cost of sending humans to, 27

    as destination, 60

    and Earth's gravity, 16

    European probes on, 173

    galactic radiation on, 84–85

    and Gateway, 242

    Indian orbiter, 177, *177*, 178

    JAXA plans to reach, 175

    journeys to, *66*

    landings on, *46*, 249, *250*

    as long-term goal for space travel, xiv, 161, 172

    moons of, 60

    orbiting of, 60, 249

    and Orion capsules, 52

    robotic rocket, *249*

    rovers on, 196, 239, *239*

    settlements on, 249–250, *253*

    simulation of bases on, 78–79

    SpaceX plans for, 125–126, 129

    successful trips to, 177

    survival on, 20, 21

    timeline to get to, 240–250

    and United Arab Emirates Space Agency, 180

    water-ice shields for living on, *91*

    water on, 206, *219*

Mars-500 simulator, 79, *79*

Mars Base Camp, *63*

Mars Exploration Rover, 196

Mars First advocates, 62–63

The Mars Institute, 270–271

Mars Orbiter Mission (MOM), 177, *177*, 178

Mars OXygen In-situ utilization Experiment (MOXIE), 239, *239*

Mars Sample Return, *249*

Mars Science Laboratory, 84

The Mars Society, 265, 266

mass, of spacecrafts, 77, 87–88

mass reduction, 122, 123, *123*

Maxar Technologies Ltd., 179

medical research, 27–28, 72–76

Mercury program, 2, *18*, 19, *43*, 44, 53, 59, 137, 178

Mercury Redstone 3 (MR-3), *43*, *44*

Merlin engine, *126*, 128, *131*

metamaterials, 107

Meyerson, Rob, 137–141, *139*

microgravity, 82, 292

Microsoft, 102, 107

militarization of space, *170*, 170–171

Mir space station, 2, 44, *46–47*, 48, 54, 73, *166*, 166–167, 184–185

Mitchell, Edgar, 32, *33*

Mojave, California, *94*

Mojave Airport, 94

Mojave Desert, 98, 111

Mojave Spaceport, *94–95*
MOM (Mars Orbiter Mission), 177,
    177, 178
moon (Earth's), 20, *158,* 161
    bases on, 247
    as candidate for human spaceflight,
        62
    capsules orbiting near, 55
    collecting resources from, *205,* 206,
        218
    as destination, 60
    exposure to radiation on, 83
    first humans on the, xiii–xiv
    galactic radiation on, 84
    and Gateway, 242
    landing on, 247
    manufacturing on, 247
    mining on, *88,* 89, 247
    north pole of, *62*
    as objective for SpaceX, 125
    orbiting of, 59, *59,* 60
    orbit of, during SELENE mission,
        175
    and Orion capsules, 52
    outpost on, *237*
    as short-term goal for space travel,
        172
    space settlements on, 242–245
    storage depots on, *221*
    surface of, 60
    travel to, *66*
    water ice on, *206*
Moon Express, 211, 218, 220
Moon First advocates, 62
moons (in general)
    of Mars, 60, 175
    Saturn's moon Titan, *170*
"Moon Village," 65, 173, 242, *243*
Moscow, Russia, *28, 79*
MOXIE (Mars OXygen In-situ
    utilization Experiment), 239, *239*
Mueller, Rob, *251*
Multi-Purpose Logistics Module, 2
Murray, Bruce, 265
muscle mass, reduction in, 73
Musk, Elon
    "billionaires' club" and, 151, 154
    Charles Bolden and, *116*
    on colonizing Mars, 34–35
    and creation of SpaceX, 115–119
    on doing something that is
        important, 1
    Falcon rockets and, 255–257, 259–
        261
    and international space agencies, 181
    Steve Jurvetson and, 158, 159
    lunar base supported by, 244
    on moving humans to space, 236
    prediction of launch prices by, 148
    and reusability, 39
    and SpaceX, 7, *118,* 123–126
    Tesla roadster, *260–261*
mutual funds, for private investors, 154
Myhrvold, Nathan, 107

N
N1 booster, 43
N1 moon rocket, *185*
NAFCOM (NASA Air Force Costing
    Methodology), 215

NanoRacks, *105,* 105–106, 154, 157, 198

nanosats, 103

NASA. *see* National Aeronautics and Space Administration

NASA Air Force Costing Methodology (NAFCOM), 215

NASA Ames Research Park, 106

NASDA (National Space Development Agency), 174

National Aeronautics and Space Administration (NASA), 290. *see also* Jet Propulsion Laboratory (JPL)

    asteroid detection systems, *224*

    and Bigelow Aerospace, 54

    budget of, 26–27, 178

    Bill Clinton's changes to, 167

    Commercial Crew Program, 135

    and cost of launching water into orbit, 88, 89

    and crewed landings, 247

    and Dream Chaser, 52, *52*

    and European Space Agency, *170*

    and Falcon Heavy launch, 257

    history of, 1–13

    and human spaceflight missions, 61, 65

    legal constraints around, 168–170

    Lunar Orbiting Platform-Gateway, *60*

    mapping of asteroids by, 224, 226, 227

    medical professionals from, 73–74

    MR-3 flight timeline from, *44*

    National Space Council and, 213, 214

    partnerships with, 9, 129–130, 140–141, 153, 158, 174, 175, 177, 178, 179, 247

    and permanent space settlement, 237–239

    psychologists/psychiatrists working at, 77–78

    and public-private partnerships, 209–211

    safety regulations at, 99–100, 107

    simulations created by, *78*

    and SLS/Orion, 52, *53,* 55

    and spaceflight-capable nuclear reactors, 90

    space infrastructure and, 209–217

    Space Portal, 199

    and space race, 42–44

    and space settlements, 237–239, 241–244

    and SpaceX Dragon, 50, *50*

    sponsored events for start-ups, *156*

    study of space colonies by, *236*

    study of space radiation by, 84

    use of robotics by, 239

    US workforce and spending by, 29–30

National Aeronautics and Space Council, 213, 214

National Defense Authorization Act (2016), 119

nationalism, 163–164

National Laboratory, 216

National Oceanic and Atmospheric
    Administration, 235
National Space Council
    defined, 290
    NASA and, 65, 213, 214
    Scott Pace and, 171, 198, 237
    reactivation of, 10, *10*
    rules of the new, 218, 220
National Space Development Agency
    (NASDA), 174
National Space Society, 266
    defined, 291
    founding of, 264–265
    International Space Development
        Conference, 35, 36
    Bruce Pittman and, 199
    Space Settlement Summit, 204
    George Whitesides and, 97
National Transportation Safety Board
    (NTSB), 99
Navy, *76, 77*
near Earth objects (NEO), 290
New Armstrong, 51
New Glenn, 51, *136,* 136–139, 244,
    245, 290–291
New Horizons Pluto mission, 132
New Line 1 vehicle, 194
New Shepard, 51, *51,* 136–139, *137,*
    139, 290–291
NewSpace, 137, 152, 291
new space race, 135–148
    Blue Origin and, 135–141
    United Launch Alliance and, 141–
        148
New Zealand, 104

normal bone density, *72*
North Korea, 180
Northrop Grumman Corporation, 102,
    121, 152, 154
Northrop Grumman Innovation
    Systems, 12, 102, 152, 174
Northwestern University, 125
Norway, 153
NSS (National Space Society), 291
NTSB (National Transportation Safety
    Board), 99
nuclear reactors, for spaceflight, 90

O

Obama, Barack, 6, 9, *56–57,* 219, 226
"offshoring," 178
OffWorld, 250
O'Neill, Gerard, 240–241, 250, 265,
    269
O'Neill space cylinders, 240–241, *241*
One Space, 194
OneWeb, 131
Orbital ATK, 12, 102, *103,* 152, *153,*
    154
orbital refueling depot, *205*
Orbital Sciences, 152, 158
orbiting stations, 63–64, 71–72
ore, from asteroids, 64
organizations, space-related, 266–271
Orion, 291
    Altair and, *8*
    capsule, *78*
    for deep-space exploration, *53*
    as government spaceflight, 246
    rendezvous with asteroid, *56–57*

Soyuz and, 186
and Space Launch System, 9, 10, 52, 55, 246, 249
Outer Space Treaty (1967), 165, *165,* 218, 219
outposts, *237–238*
overview effect, 32, *32*
Oxford University, 235
ozone layer, 228

P

Pace, Scott, 171–172, 196, 198, 214, 237
Paleogene era, 223
partial-gravity environments, 22, 81
Passengers (movie), 81
passive radiation shielding, 90
payloads, 119, 126, 194, *259, 261,* 291. *see also* fairings
Pegasus rocket, 152–153
Pegasus satellite, 102, *103*
Pence, Mike, 214, 220
People's Liberation Army (PLA), 193
Pew Research Center, 252
Phobos, 60, 175
Physics Today (journal), 240
Pickering, Bill, *41*
Pittman, Bruce, 199–200, 205–206, 208
PLA (People 's Liberation Army), 193
Planetary Resources, 158, 180, 217
The Planetary Society, 230, 265, 266
Planet Labs, *105–106,* 106, 159, 160, *160*
plasma shielding, 90
plastic, in spacecrafts, 87

Pluto mission, 132
pressure suits, 69–71, *70, 199*
pressurized, 291
Prince, 34
Princeton University, 97, 265
private investment opportunities, 152–155, *157*
private spaceflight, 291
probes, 173
propellant, 291
PSLC rocket, *176*
psychiatrists, at NASA, 77–78
psychological issues, from confinement, *76,* 79
psychologists, at NASA, 77–78
public-private partnerships, 209–211
Putin, Vladimir, 186

Q

Quayle, Dan, 214

R

RAD (Radiation Assessment Detector), 84, *85*
radiation, *21,* 21–22, 82–91
    genetic damage from, 86
    map of solar, *84*
    shielding from, 90–91, *91*
    sources of, 83–84
    Van Allen belts as defense against, *83*
Radiation Assessment Detector (RAD), 84, *85*
RD-180 rocket engine, *149*
Reagan, Ronald, 167, 221

"Recommendations to the Next Administration Regarding Commercial Space," 213–217

Red Army, 192

Red Planet. *see* Mars

reentering the atmosphere, 18, *19*

regolith, *89*

research and development (R&D), 217

resources
    collecting, *205*
    competition over, 218–221
    extraction of, *158*, 218, 234, 247
    storing mined, *88*
    utilizing, for space settlements, 234, 235

retrorockets, 18, *18*

reusability
    Blue Origin and, 197
    defined, 291
    SpaceX and, 39, 119–121, *125*, 126–128, 204, 259
    United Launch Alliance and, 143–145

Richter scale, 227

robotic landings, 247

robotic probes, *64*

Rocket Lab, 103–104

rockets, 103, 104, 122–123, *131*. *see also specific rockets, e.g.:* Saturn V

Rohrabacher, Dana, 238

Roscosmos, 186–187, 206, 291

Rosen, Stanley, 32, 235, 236

Russian Academy of Sciences, 79

Russian space program, xiv, 12, 183–188. *see also* Soviet Union

achievements of, 183–184

challenges facing, 184–187

Chelyabinsk asteroid collision, 226

collaboration with India, 177–178

cooperative crewed missions with US, 167

crewed stations in LEO, 61–62

future of, 187

international collaboration, 48

launch vehicles, 186, 187

lowering launch costs, 148

Mars aspirations of, 65, 79

militarizing space, 170–171, 220–221

Mir space station, 2, 44, *46–47*, 48, 54, 166–167, *167*, 184–185

NASA partnership with, 247

nuclear weapons against asteroids, 230

radiation guidelines of, 86

selling parts to United Launch Alliance, 140

Soyuz, 7–8, 42–43, 44, *47*, 48, 49, *49*, 55, 59, 183, *185, 189*, 195, *201*, 246

space entrepreneurs in, 181

and Ukraine, 153

Zarya module, *171*

Rutan, Burt, 94, 99, 102

S

SAA (Space Act Agreements), 215

safety, of spaceflight systems, 99–100, 107

Sagan, Carl, 265

Saint-Exupéry, Antoine de, 1

Salyut space station, 71–72, 184

satellites

 Chinese, 195–196

 commercial, 28–29

 global broadband internet via, 130–132

 Indian, 176, 177, 178

 in low Earth orbit, 107

 Planet Labs, *160*

 for small launches, 102

 solar-power, 204, 234, 248

 space entrepreneurs and, 93–100

 weather, *235*

Saturn, *170,* 197

Saturn V moon rocket, *17,* 44, *45,* 52, 115, 128, 129, 216, 255, 256

Saudi Arabia, 155, 180

Scaled Composite, 99

Sea Launch, 153, 154

SEDS (Students for the Exploration and Development of Space), 269

SELENE mission, 175

The Seti Institute, 268

settlement model, expeditionary vs., 237

settlements in space. *see* space settlements

747 carrier plane, *101*

Shackleton, Ernest, 233

Shanghai Science and Technology Museum, *49, 201*

Shedd, John A., 99

Shenzhou stations, 49, *49,* 193, *193,* 195, *195, 199, 201,* 291

Shepard, Alan, 42, *43–44,* 136

Shotwell, Gwynne, 117–118, 121, *124, 124*–126, 128–131

Shuguang spacecraft, 192

Siddiqui, Asif, 186–187

Sierra Nevada Corporation, 12, 52

Silicon Valley, 31, 211

Sino-US partnership, 195

65803 Didymos (asteroid), *228,* 229

Skylab, *46,* 71–72, *72,* 77, *87,* 256

SLC-4E (Space Launch Complex 4E), 112

sleep cycles, 76

SLS. *see* Space Launch System

smallsats, 103

smartphones, 105, 106

Smithsonian Institution, 236

solar-power satellites, 204, 234, 248

solar radiation, *84*

South Korea, 180

Soviet Union. *see also* Russian space program

 Energia booster, 128

 "handshake in space," *166*

 international cooperation with, *164,* 198

 Mir space station, 44, *167*

 N1 moon rocket, *185*

 Salyut, 71

 and space race, 42, 69, 163–165, 176, 194

 Sputnik, 40, *40,* 183, 191

 unsuccessful trips to Mars, 177

Soviet Venera 13 probe, *184*

Sowers, George, 146–147, 208

Soyuz, 291–292

Apollo docking with, *166*

Apollo-Soyuz Test Project, 165–166

cost of seats on, 8

for government spaceflight, 246

for human travel, 49

improvements to, from Chinese
    National Space Agency, 195, *201*

and International Space Station, 48,
    48, 49, 185, 189

landing, *19*

in low Earth orbit, 59

replacement of, 185–186

for travel to the moon, 42–43, 55,
    183

space

    aurora as seen from, *61*

    commercialization of, 212–217

    defined, 292

    involvement in, 263–271

    medical impact of, on the body,
        72–76

    militarization of, *170,* 170–171

    sunset as seen from, *16*

    and survival of humanity, 20–21,
        34–36

    temperatures in, 21

    travel destinations in, 59–60

    travel to and from, 15–23

Space 2.0, 292

Space Act Agreements (SAA), 215

space age

    cooperation between US and Russia
        during, 48

    first, 39–55

    and human-rated spacecrafts, 49–55

rebirth of, 255–261

    and space race, 40–44, 48

space agencies, international, 180–181

Space Angels, *155,* 155–158, 220

space colonies, *236*

spacecraft. *See also specific spacecraft, e.g.:*
    Soyuz

    Chinese, 192

    human-rated, 49–53

    Japanese, 175

    mass of, 77, 87–88

    materials comprising, 87, 89

space entrepreneurs, 93–108. *see also*
    space investors

    Jeff Bezos, 101–102

    Blue Origin, 101–102

    Richard Branson, 94, *97,* 100

    defined, *292*

    Firefly Aerospace, 103

    Kymeta, 102

    Made In Space, Inc., 106–107

    NanoRacks, 105, 106

    Planet Labs, 106

    Rocket Lab, 103–104

    and space settlements, 236

    Stratolaunch, 102

    Vector Space Systems, 104

    Virgin Galactic, 93–101

    Virgin Orbit, 100–102

    George Whitesides, 97, *98,* 100–102

space exploration, 25–36, 234

Space Exploration, Development, and
    Settlement Act (2016), 238–239

Space Exploration Technologies Corp. *see*
    SpaceX

spaceflight
    destinations for, 59–60 (*see also*
        specific destinations, e.g.: Mars)
    government, 246
    and human factors, 44–48
    Indian crewed flights, 177–178
    investors' interest in, 158
    and nuclear power, 90
    private, 291
    safety considerations in, 99–100, 107
Spaceflight magazine, 268
The Space Foundation, 267–268
Space Frontier Foundation, 268
Space Generation Advisory Council, 271
Spacehub, 153, *153*
space infrastructure, 203–221
    Aldrin Cyclers, 211–212, *213*
    and commercialization of space,
        212–217
    and competition over resources,
        218–221
    creating, 204–206
    and deep-space economy, 206, *207,*
        208–211
    defined, 292
    elements of, 204
    groups developing, 217–218
    NASA and, 209–217
    National Space Council, 213, 214
space investors, 151–161. *See also* space
    entrepreneurs
    Chad Anderson, 155–158
    ATK, 152–153
    Jeff Bezos, 151, 154
    Steve Jurvetson, 155, 158–160, *159*

Elon Musk, 154, 158, 159
mutual funds for, 154
NanoRacks, 154, 157
NewSpace, 152
Orbital Sciences, 152, 158
Planetary Resources, 158
Planet Labs, 159, 160, *160*
Sea Launch, 153–154
Space Angels, *155,* 155–158
Spacehub, 153, *153*
Space Services Incorporated, 152,
    153, 154
SpaceX, 156–160
Tesla, Inc., 159
Space Launch Complex 4E (SLC-4E),
    112
Space Launch System (SLS)
    Block 1 shuttle, *53*
    and Blue Origin, 140, 244
    defined, 291
    Gateway launched by, 242
    Orion and, 9, 10, 52, 55, 196, 246,
        249
Space Medicine Association, 74
Space Policy Institute, 186–187
Space Poop Challenge, 107
Space Portal, 199
space race. *see also* new space race
    defined, 292
    and international cooperation,
        163–168
    in 1960s, 13, 42–44, 69
Space Services Incorporated (SSI), 152,
    153, 154
space settlement, 62–65, 233–252, 248

creating incentives for, 250–252

defined, 292

entrepreneurs and, 236

and human survival, 234–236, 238

on Mars, 249–250

on the moon, 62–63, 242–245

NASA and, 237–239, 241–244

non-human, 235

and resource utilization, 234, 235

technology to support, 239–242

timeline for, 245–250

via O'Neill space cylinders, 240–241, *241*

Space Settlement Summit (2017), 204, 205

SpaceShipOne, 94

SpaceShipTwo, 98, 99

space shuttle. *see* US space shuttle

Space Studies Institute, 240, 241, 269

space suits, 69–71, *199*

Space Symposium, 267

Space Systems/Loral (SSL), 179

Space Tech Summit 2018, 161

space tourism, 59, 101–102, 137, 245, 264

Space Transportation System, 2

space tugs, *145,* 146, 148

spacewalks, 70, *71,* 184

SpaceX, 111–132, 156–160

    Big Falcon Rocket, *86*

    Jim Cantrell and, 104

    Commercial Crew Program funding, 135

    as commercial spacecraft, 7, 9–12, 194, 245

and cost of launching water into orbit, 88, 89

and crewed landings, 247

defined, 292

and Dragon rockets, 19, 50, *50,* 119, *123,* 126, *159,* 185, *210*

Falcon 9, 215, *215,* 255–257

Falcon Heavy, *146*

and global broadband internet access, 130–132

headquarters, *121,* 121–122, *124*

investors of, 129–130

and isogrip machining, *123*

landings on Mars by, 249

large rocket development by, 52, 55

late deadlines of, 187

and launch costs, 131

launch sites, 111–113, *112–118, 120–121*

in low Earth orbit, xiv

Elon Musk and, 115–120, *116*

NASA partnership with, 129–130, 140, 174

pilot space settlements by, 248

plans for Mars, *125,* 125–126, 129, 244

public support of, 252

and reusability, 39, 119–120, *125,* 126–128, 186, 197, 204

rocket factory, *122,* 122–124, *126, 159, 256*

satellites launched by, 102

Gwynne Shotwell and, 124–126

Tesla roadster, *261*

United Launch Alliance vs., 142–145

Virgin Galactic vs., 121
Spudis, Paul, 233
Sputnik, 40, *40,* 183, 191
SSI (Space Services Incorporated), 152, 153, 154
SSL (Space Systems/Loral), 179
Stafford, Tom, *166*
Stanford Torus, 241
Starbucks, 29
star cluster, *83*
Starliner, 11, *12,* 50, *50,* 55, 185
"Starman," 260, 261
stars, as source of radiation, 83–84
"Starship." *see* Big Falcon Spaceship
Star Trek (television), 80, 233
STEAM activities, 266
STEM activities, 266
Stern, Alan, 132
storage depots, 208, *221*
storm shelters, 90
Stratolaunch rocket, 102, 245–246
Students for the Exploration and Development of Space (SEDS), 269
Styx, 34
submarines, *76,* 76–77
suborbital flight, 245
suits, for astronauts, 69–71
sun, as source of radiation, 83–84
sunset, from space, *16*
superconductors, 90
"Super Heavy." *see* Big Falcon Rocket
survival of humanity, 20–21, 34–36

T
"taikonauts," 192, 193

tankers, *88*
Tarver, Bill, 73–76, 78–79, 85
Tau Zero Foundation, 267
temperatures
    impact of asteroids on, 228
    in space, 21, 70
territoriality, 76–77
Tesla, Inc., 159
Tesla roadster, 260, *260–261*
Teslar satellite, 211
3-D printers, 106–107, 206, *211,* 242–243, 247
3U cubesats, 106
Tiangong space station, 12, 54, 193, *193,* 195, 196, 291
tiles, heat-absorbing, 18–19
timeline, for space settlement
    cislunar Space, 246–248
    Earth orbit, 245–246
    Mars, 249–250
    suborbital fight, 245
TOPEX Poseidon ocean observing system, 169
tourism, in space, 59, 101–102, 137, 245, 264
Trump, Donald, 10, *10,* 171, 220–221
Tsiolkovsky, Konstantin, 16
tsunamis, 227
Tumlin, Rick, 218–219
twin boosters, *259*
2001: A Space Odyssey (movie), 81

U
UAESA (United Arab Emirates Space Agency), 180

Ukraine, 153

ULA. *see* United Launch Alliance

UN (United Nations), 165, 271

United Arab Emirates, 180

United Arab Emirates Space Agency (UAESA), 180

United Launch Alliance (ULA), *119*

    ACES stage, *145*

    Advanced Programs, 147

    Atlas rockets, 11, 118, 122, *142, 164*, 175, 187

    Atlas V rocket, 50, 55, *119*, 135, 140, 142, 143, *143, 147, 149*, 153, 188, 289

    Tory Bruno and, *143*

    CisLunar-1000 Economy, *207*, 208, 214–215

    Commercial Crew Program funding, 135

    defined, 292

    Delta IV Heavy, 142, *146*

    Delta IV rocket, 52, 55, 117–119, 135, 142, *146*, 289

    Delta rockets, 118, 122

    headquarters, *141*

    in low Earth orbit, xiv

    NASA partnership with, 140–141

    and new space race, 141–148

    part purchases from Russia, 140

    and reusability, 204

    as rival to Blue Origin, 137

    rockets of, 55

    SpaceX vs., 142–145

    Starliner, 50

    US Air Force and, 117–119, *119*

Vulcan rocket, 135, 137, 142–143, *144*, 246

United Nations (UN), 165, 271

Unity spacecraft, 51, 94–97, *95*, 100, *100–101*, 101, *171*

University of Houston, 137

University of Michigan, 137

University of Rochester Medical Center, 86

University of Strathclyde, 230

University of Tokyo, 174

US Air Force

    as branch of US armed services, 221

    Tory Bruno and, *143*

    Falcon 9 rocket, *133*

    Stanley Rosen and, 235

    SpaceX and, 114–115, *118–119*, 130

    United Launch Alliance and, 112, 117–119

US Army, 221

USB-C connector, 105, 106

US Coast Guard, 221

US Congress, 25, 195, 264, 265

US Department of Commerce, 29

US Department of Defense, 119

US House of Representatives, 238

US Marines, 221

US Navy, 221

US Space Force, 171, 221

US space shuttle, *16, 46–47*, 292

V

Van Allen, James, *41*

Van Allen belts, 83, *83*

Vandenberg Air Force Base, 111, *112–115, 118, 133*

van Zyl, Jakob, 34, 198

Vector Space Systems, 104, *104*

Venera probes, 183–184

Venus, 20, 21, 177, 183–184, *184*

Viking 1 spacecraft, *46, 250*

Virgin Galactic, *95–98*

    in Mojave Airport, 94–95

    and Saudi Arabia, 155

    SpaceX vs., 121

    tourist flights offered by, 59, 245

    Unity and, 51, *100,* 100–108, 137

Virgin Orbit, 100–102, *101,* 245

von Braun, Wernher, *40, 41,* 265

Voskhod, 55

Vostok, 55, 184

Vulcan rocket, 135, 137, 142–143, *144,* 246, 289

W

Washington Post, 238

waste, elimination of human, 70, 107

water, 62–63, 64, 88, *206, 219,* 247

weather satellites, *235*

weightlessness, 72–75, *73,* 80, 86, 292

Weitz, Paul, *72*

Western Test Range, 111

White, Ed, *71*

White, Frank, 32

WhiteKnightTwo, 96

Whitesides, George T., 97–98, *98,* 100–102

windows, in space stations, *77*

work shifts, 76

World War II, *76,* 176

Wörner, Johann-Dietrich, 173, 242–243

X

X-15 rocketplane, 59, *70,* 95, 96, *96*

XARM (X-Ray Astronomy Recovery Mission), 175

XEUS lunar lander, 208–209

X-Ray Astronomy Recovery Mission (XARM), 175

Y

Yonsei University, 170

Young, John, *71*

Yucatán Peninsula, 224, *225*

Yuri's Night, *97, 270*

Z

Zarya module, *171*

Zeitlin, Cary, 85

zero-g, 292

Zond spacecraft, 184

Zubrin, Robert, 30–31, 35, 168, 265

ABOUT THE AUTHOR

Rod Pyle is a space author, journalist, and historian who has written thirteen books on space history, exploration, and development for major publishers that have been published in seven languages. He is the senior editor of *Ad Astra* magazine, the quarterly print publication of the National Space Society, and his articles have frequently appeared in *Space.com, LiveScience, Futurity, Huffington Post*, and *WIRED*.

Pyle has written extensively for NASA's Jet Propulsion Laboratory and Caltech, and authored the *Apollo Executive Leadership Program* for the Johnson Space Center with The Conference Board. His most recent book release is *Amazing Stories of the Space Age* (Prometheus Books, 2017) and is currently under option for television. Pyle writes a yearly technical publication called *Technology Highlights* for NASA/JPL.

He appears frequently on national radio and television, with regular slots on KFI/Los Angeles and WGN/Chicago. He was recently featured on PBS's "Between the Lines" and C-SPAN's "Book TV." Pyle holds an MA from Stanford University and a BFA from the Art Center College of Design.

He lives in Alhambra, California.

Photo by Geoffrey Notkin

JOIN THE NATIONAL SPACE SOCIETY

Your ticket to a spacefaring civilization—join the NSS today, and make a difference.

1. **Advance the Movement** . . . Become part of a **dynamic community** that fosters the exploration and settlement of space.
2. **Stay Informed** . . . *Ad Astra*, the award-winning magazine and the *Ad Astra* **Downlink** e-newsletter keep you up-to-date on NSS activities and the latest advances in space settlement.
3. **Get Energized** . . . Attend the **International Space Development Conference (ISDC®)** and connect with space leaders who will share the latest breakthroughs in space exploration and development.
4. **Shape Space Legislation** . . . Be Heard and *"Make the Case for Space"* through the **NSS legislative** and **Advocacy programs**.
5. **Engage, Motivate, and Inspire** . . . See your ideas come to life through NSS **educational initiatives** and **projects**.
6. **Share Your Passion for Space** . . . Meet like-minded advocates for space through **NSS chapters** and programs like the **Space Ambassadors**.
7. **Lead the Way** . . . **Become an NSS Volunteer** on the local or national level and take part in tangible activities that move the space development agenda forward.
8. **Get Rewarded** . . . NSS membership includes an invitation to join the **NASA Federal Credit Union** where you can enjoy competitive rates on credit cards, loans, and deposits.

For more information on how to join the NSS, go to **space.nss.org**, or call (202) 429-1600.